THE HANCOCKS OF MARLBOROUGH

The Hancocks of Marlborough

Rubber, Art and the Industrial Revolution: A Family of Inventive Genius

JOHN LOADMAN
and
FRANCIS JAMES

OXFORD
UNIVERSITY PRESS

Great Clarendon Street, Oxford OX2 6DP

Oxford University Press is a department of the University of Oxford.
It furthers the University's objective of excellence in research, scholarship, and education by publishing worldwide in

Oxford New York

Auckland Cape Town Dar es Salaam Hong Kong Karachi
Kuala Lumpur Madrid Melbourne Mexico City Nairobi
New Delhi Shanghai Taipei Toronto

With offices in

Argentina Austria Brazil Chile Czech Republic France Greece
Guatemala Hungary Italy Japan Poland Portugal Singapore
South Korea Switzerland Thailand Turkey Ukraine Vietnam

Oxford is a registered trade mark of Oxford University Press
in the UK and in certain other countries

Published in the United States
by Oxford University Press Inc., New York

First published 2010

British Library Cataloguing in Publication Data
Data available

Library of Congress Cataloging in Publication Data
Data available

Typeset by Newgen Imaging Systems (P) Ltd., Chennai, India
Printed in Great Britain
on acid-free paper by the
MPG Books Group, Bodmin and King's Lynn

ISBN 978–0–19–957355–4

1 3 5 7 9 10 8 6 4 2

This book is dedicated to our true friends who have stood by us and not run a mile whenever we started talking about the Hancock family—or rubber—and particularly Frank's brother Professor Tom James MA PhD FSA with whom he has shared the Hancock adventures over many years, and whose wise counsel has been invaluable.

The Lord Gave, and
the Lord
Hath Taken Away
Job 1: 20–21 (King James Bible)

Contents

Preface

The story of Thomas Hancock's life has never been told. In his own day he was acknowledged as one of the great scientific pioneers of the Industrial Revolution, yet his reputation has been allowed to fade like so many old photographs which have been exposed to the sun until they are almost invisible.

The role he played in the progress and development of the rubber industry was unequalled. Not only did he discover the means of transforming rubber into a useable material and invent the machinery with which to do this, he was also the first person to develop a complete understanding of the process by which rubber could be treated with sulphur to cure it of the two properties which rendered it all but useless for many applications: the fact that it softened and flowed when warm but yet became as brittle as a potato crisp or as hard as a block of modern plastic when cold. His patent of 21 November 1843 must give us one of the most important dates in the history of the Industrial Revolution.

By developing both the vulcanization process and the processing equipment with which to form and shape his products he did not merely found an industry, he created and constructed a completely new one which opened the floodgates for companies worldwide to manufacture innumerable products which it had previously been impossible even to conceive without the benefit of this unique elastic material.

How has it happened that two people have come together to chronicle not just his most interesting life but also the family saga wherein the lives of his siblings were inextricably interwoven with his own and where, after his death, the family continued for a further two generations to produce men of character and ability who successfully continued his business?

Thomas Hancock would have said it was due to the mysterious hand of Providence, and it would be a brave man who would dare to contradict him.

There have been two major researchers into the history of rubber. In the early part of the last century Benjamin D. Porritt, Director of Research at the British Rubber & Tyre Manufacturers Research Association, was indefatigable in tracing many of the original documents and ephemera from the early days of the industry and for ensuring that they were safely preserved for posterity. John Loadman has played the same role in continuing Porritt's researches and in the course of his own research came upon material that Porritt collected from the Hancock family, but which had lain hidden and ignored after his death in a box buried deep in the archives of the Rubber and Plastics Research Association (Rapra). The directors kindly agreed to return it to the family archivist, Francis James, a lineal descendant of Thomas' brother John Hancock, who has spent a lifetime assembling the Hancock Archive and has an extensive knowledge of the more personal details of the lives of this extraordinary family.

Had it not been for the discovery by John Loadman that the archivist of the Plastics Historical Society (PHS) had recovered the Hancock family bible from a skip and John's suggestion that this too belonged in the Hancock Archives, something to which the PHS Committee immediately agreed, your two authors might never have met, but when this did occur at the handing over of the bible to the Hancock family, we discovered that we had both been thinking that it was high time this biography was written. With much of the information that we had being complementary, collaboration rather than conflict seemed the perfect way forward.

As we pooled records, documents and correspondence, it soon became apparent that there was a fascinating story to be told which extended both before and after the time Thomas spent on this earth and the simple biography grew into the rise and fall of the Hancock family rubber fortunes: a period close to 120 years.

Thomas Hancock, born at the time of the French Revolution and growing up through the Napoleonic Wars, was driven by two ruling principles: a strong sense of duty to both his God and to mankind. He was truly a Renaissance Man in whom the wiles of Machiavelli were coupled with the inventiveness of Leonardo da Vinci, but coloured by the uncompromising fierceness of Calvin. He was a complicated character, austere and possessed by an unwavering devotion to truth in everything he did, both in his private life and in all his business dealings. He was certainly shrewd and secretive, and showed considerable cunning in protecting the interests of himself and his family, yet at the same time he could not resist the innocent charm of children with whom he was always a great favourite; with over 20 nephews and nieces vying for his attention he certainly had plenty of opportunity to try!

He instilled the same values into his nephew and close assistant, James Lyne Hancock, and we shall show how James continued the family tradition by successfully expanding Thomas' original company to the end of his life. At his death in 1884, he willed it to the next generation of the family in the shape of John Hancock Nunn. It is a sad but inevitable fact that few families can produce great business-men or entrepreneurs indefinitely and John Hancock Nunn's demise in 1937 led to the end of Thomas' family business as an independent company.

Reclusive by nature, Thomas wrote very little about his personal life, although in later life he did review his part in the progress of the rubber industry. He seems to have made himself invisible, so that we may know him only by his works, and he would have thought this exactly as it should be. Even the town of his birth, Marlborough in Wiltshire, has never shown the slightest interest in according him the reputation he deserves. Mr Porritt found this when he visited Marlborough in the early years of the last century and was shocked to discover that the name of Thomas Hancock was unknown; this hardly more than 40 years since his death! Francis James found that nothing had changed in this lack of appreciation when he moved to Marlborough in 1974 to open an antiquarian bookshop and pursue his Hancock family researches. When, belatedly, Marlborough decided a few years ago to give a slight nod of recognition to its famous son by the erection of a commemorative blue plaque, it was mounted on the wrong building!

How has this great man become so invisible? In modern terms, we could say that he lost brand name recognition. When he subsumed his interests into the Charles Macintosh business and, many years before his death, passed his original business into the capable hands of his nephew, James Lyne Hancock, his name vanished from the public eye. Although the eponymous garment that bore the name Mac(k)intosh became a household word, Hancock's association was only discernible by the Macintosh and Co. trademark, a hand on which is perched a cockerel, discreetly sewn into the lining.

While Thomas was the only one of the six brothers to make a considerable amount of money during his lifetime and, indeed, often offered financial assistance to his siblings, this generation of the family must be unique in the range of their achievements. Thomas' younger brother, John, began his own rubber business, specializing in medical applications and died young—leaving Thomas to look after his family of nine children. Next came William, who was a master cabinet-maker and who has at least one piece in the royal collection of Queen Elizabeth II. Walter was a pioneer of the steam carriage age, designing, building, and running the first commercial steam carriage or omnibus service in London while Charles was a famous and talented artist who exhibited regularly at the Royal Academy, the Royal Society of British Artists, and the British Institution. It is to him that we owe many of the likenesses of his family. He was also one of the first to realize the significance of the newly discovered gutta percha as an insulant for electric telegraph cables and started a company to manufacture a range of gutta percha articles. Only a fatal character flaw prevented him from reaching heights comparable to those achieved by Thomas.

The following generations produced their own characters, although not on the scale of Thomas and his siblings.

This 'saga' examines the lives of many of these characters as they wove in and out of the fabric of Thomas' story and the lives of the rubber companies with which he was involved.

John Loadman and Francis James

2009

Acknowledgements

Although both of us had collections of information about Thomas Hancock and his family, the true genesis of this book lay in our search for a collection of over 700 lantern slides for which we had a catalogue dating from around 1926. It was obvious that the first 100 or so of these related to the Hancock family.

It was during this search that we contacted Jackie McCarthy at the Rubber and Plastics Research Association (Rapra), who, while searching unsuccessfully for the slides with Sheila Cheese, came across a dust-covered box containing numerous documents and correspondence relating to, or written by, Thomas Hancock which had been 'loaned' to the forerunner of Rapra in the early years of the twentieth century and had lain 'lost' ever since. Our thanks are extended to Carole Lee at Rapra for returning the documents to the family archives. One of the documents was an early draft of a talk which was given by Messrs Porritt and Dawson to the Institute of the Rubber Industry in 1924, and it was obvious that the slides referenced in the talk correlated with those at the beginning of the catalogue list.

It took longer to locate the lantern slides themselves and to discover their history. They were accumulated and catalogued in the 1920s by the Research Association of British Rubber Manufacturers, where they remained virtually unknown and forgotten until they were about to be thrown away in the 1990s. Luckily the archivist of the Plastics Historical Society (PHS), Colin Williamson, saved them and they were donated to the PHS. Acknowledgement is given to the PHS for the use of the collection and most particularly to Colin for scanning many of the slides on to compact disks for us. Inevitably this provides a quality which does not compare with modern photographs but they occupy a unique place in the history of British rubber industrialization and cannot remain unpublished. Figs 7.4 and 7.5 were also supplied by Colin Williamson and are used with permission. The return of the family bible has been mentioned in the introduction and for this it is again most appropriate to thank Colin and the PHS.

Other illustrations have a range of histories with many being from the family archives. Indeed, there is a degree of duplication between those and the lantern slides which suggests a close cooperation between

Mr Porritt and the family in the early part of the twentieth century. For that reason individual credits are not given. The letter from Thomas Hancock to Sir William Hooker, Director of Kew Botanic Gardens, from which extracts are quoted in Chapter 10, was initially located in the 'Miscellaneous Correspondence Collection' of the Royal Botanic Gardens, Kew, by Hew Prendergast, and was supplied to us by Michele Losse and Julia Steele. The photograph of the unusual shoe, Fig. 10.9, was supplied by Sue Seddon for an article in the *Kew Magazine*. These are published with the permission of the Trustees of the Royal Botanic Gardens. Fig. 4.2, William Hancock's elm cabinet, is from the Royal Collection © Her Majesty Queen Elizabeth II, and is reproduced with permission. Our thanks go to David Oakey and Katie Holyoak of the Royal Collection for facilitating our use of it. The articles relating to this cabinet, which are reproduced from the *Bury and Norwich Post*, were found in the Suffolk Record Office in Bury St Edmonds and their permission to use them is acknowledged with thanks.

Our thanks are also due to Ted Rogers of the Hackney Borough Archives, who helped us trace the location of Thomas Hancock's home, Marlborough Cottage, from census and land registry records, although its well-established name was not found in any of the documentation. This enabled the PHS to place a plaque on the building currently occupying the site of Thomas' home and laboratory.

Information on the Gutta Percha Company from which Figs 8.2–8.4 were taken, was found in the archives of East Sussex and the figures are reproduced with the permission of the County Archivist of East Sussex—copyright reserved. Fig. 8.11 was supplied by an American friend, Mike Woshner, and his permission to use it is gratefully acknowledged.

The map of Stoke Newington in 1848 (Fig. 5.1) was taken from the website of British History Online. Attempts to trace any copyright restrictions on our use of it were unsuccessful. British History Online referred us to Landmark Publications from whom they obtained it and whom, they said, held the copyright. Rick Crowhurst, Data Sales Manager of that company, is thanked for researching the map and deciding that it was not actually covered by any copyright. If this or any other illustration has 'sneaked through' without being appropriately accredited, we hope that we shall be forgiven and the original owner will be content to see his or her work published here.

Second-hand bookshops and internet searches for referenced books provided more information, as did the library and archives of the Tun

Abdul Razak Research Centre near Hertford, where Gail Reader is thanked for her help in locating many books and documents of background interest. Sir Alex Moulton is thanked for providing us with a copy of the privately published book, *A Hundred Years of Rubber Manufacture 1848–1948,* the story of Stephen Moulton and George Spencer Moulton and Co. Ltd.

While the disposal of books and journals such as the *India Rubber Journal* and *India Rubber World* by universities which had no space in their libraries (or syllabuses!) for old science books and their contents can do little to further the general education of their students, they provided a fund of information for the latter part of the book for which we are grateful.

Roger Street is thanked for providing additional information about the Pilentum (Fig. 2.1), while Brigadier Antony Karslake is thanked for providing a photograph of the testimonial (or mausoleum!) (Fig. 12.5) which Thomas gave to his trusted friend and solicitor, Mr Henry Karslake. Finally the contribution Katherine Ingram, a descendant of William Hancock, made in helping Francis organize the family archives is noted with appreciation, while thanks are also due to members of Francis' long-suffering extended family for their input, with particular reference to Professor Tom James, while our respective wives, Lina and Janet, are thanked for their tolerance as well as for reading and commenting on the book.

List of Illustrations

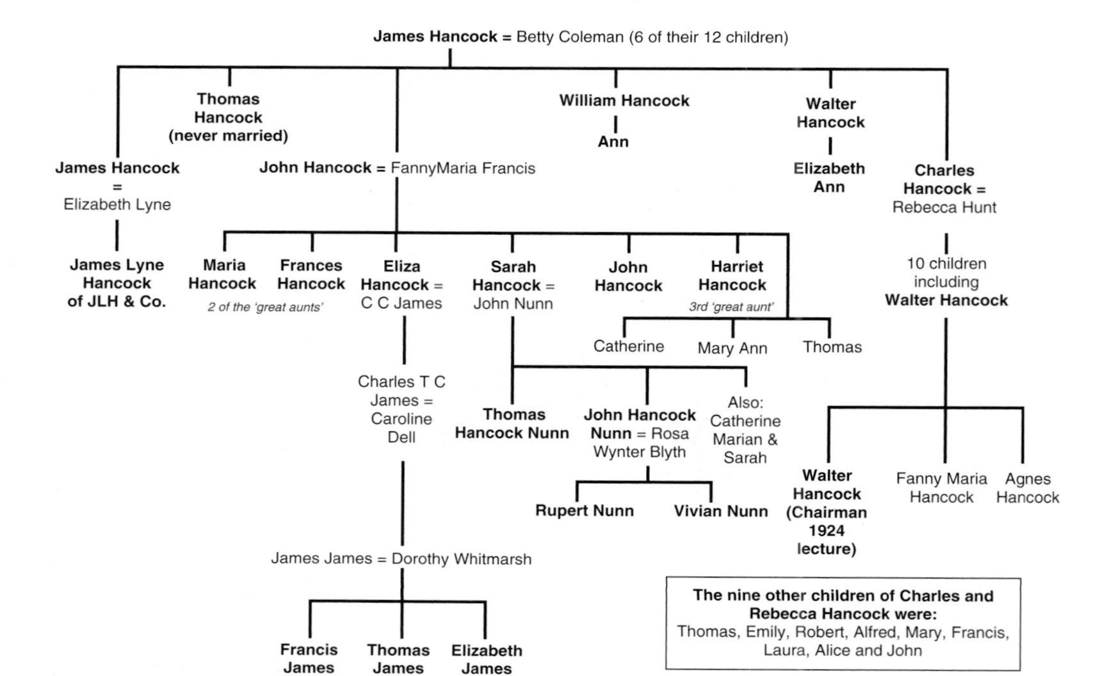
The Major Players in the Story of Thomas Hancock
and his Rubber Companies
James Hancock = Betty Coleman (6 of their 12 children)
James Hancock = Elizabeth Lyne
Thomas Hancock (never married)
John Hancock = FannyMaria Francis
William Hancock
Ann
Walter Hancock
Elizabeth Ann
Charles Hancock = Rebecca Hunt
James Lyne Hancock of JLH & Co.
Maria Hancock
Frances Hancock
2 of the 'great aunts'
Eliza Hancock = C C James
Sarah Hancock = John Nunn
John Hancock
Harriet Hancock
3rd 'great aunt'
Catherine
Mary Ann
Thomas
10 children including Walter Hancock
Charles T C James = Caroline Dell
Thomas Hancock Nunn
John Hancock Nunn = Rosa Wynter Blyth
Also: Catherine Marian & Sarah
Rupert Nunn
Vivian Nunn
Walter Hancock (Chairman 1924 lecture)
Fanny Maria Hancock
Agnes Hancock
James James = Dorothy Whitmarsh
Francis James
Thomas James
Elizabeth James
3 great great great nephews/niece at the unveiling of the plaque to Thomas Hancock
The nine other children of Charles and Rebecca Hancock were:
Thomas, Emily, Robert, Alfred, Mary, Francis, Laura, Alice and John

1

Marlborough, Wiltshire—Roots

1653, the Great Fire of Marlborough—influx of craftsmen including the Hancock family—1753, birth of James Hancock—marriage to Betty Coleman and growth of their family (12 children)—importance of education—James' business as a craftsman and enrolment of his sons (and one daughter) as apprentices. Growth of Marlborough—the family bible and secrets contained therein. Eruption of Mount Tambora and climatic effect on the UK—retirement and death of James and Betty Hancock—migration of the brothers, James junior, Thomas, John, William, Walter, and Charles to London.

The pre-history of the Hancock dynasty of artists, craftsmen, and scientists in London in the nineteenth century began in the wake of a local disaster in a small market town in Wiltshire—the Great Fire of Marlborough of 28 April 1653. It was probably this disaster that brought the family to that town to take part in its rebuilding.

By coincidence it was following a global disaster, the eruption of Mount Tambora in 1815, that on 28 April 1818, almost exactly 165 years after the fire, the Hancock cabinet-making and carpentry business in the town drew to a close with a sale of all the stock and equipment. While the Hancock story opens locally in Wiltshire, it will erupt on to the national and international stages from 1818 onwards as the family members display their singular and diverse talents.

The Marlborough to which they migrated in 1653 is vividly described in a pamphlet of great rarity[1] issued a few days after the fire, which sets the scene for the town in which the family was to develop over the next century and a half or so.

The famous and flourishing town of Marlborough in Wiltshire had of late two faire parish churches, one called by the name of St Peter's and the other church called by the name of St Mairie's; there was likewise many faire streets and stately buildings, especially one gallant street called the High Street, in which they kept their markets, which markets consisted of all kinds of necessarie provisions, which was brought in from far and near by the country people. And indeed it was a gallant place for Corn, Butter, Cheese and such like provisions, as any was in all the country. The Street wherein the market was kept, is supposed to be in length and breadth full as large as Cheapside, and on both sides had many goodly shops, well fitted with rich and costly commodities, silks and tafety cloaths and lace linen and woollen, gold and silver, no braver wares can be had or bought in London than was to be had in the famous towne of Marlborough.

At the upper end of the Market Place was a gallant building called the Town Hall, wherein the magistrates sat and held the sessions of the peace at the appointed times; there were many faire inns, taverns and victualing houses, to entertain carriers and travellers and such which had occasion to make use of them, for it stood upon the roade between London and Bristoll; and to be brief it was a towne of very good orders and government...

On Thursday 28 April in the house of one Mr Freeman, a tanner, as some of his servants were imployed with drying of barke, the barke took fire so suddenly that it quickly did much harme, the house standing on the south side of the streete towards the west end of the towne, near unto St Peter's church, the fire prevailed so much that it tooke hold of the dwelling house, and so running acrosse the street from one side to the other, it came to be of such force and vehemency that the like was never seen in England before, by the report of some of them who were witnesses of that sad object.

It burned in both sides of the street all the inns, taverns, gentlemen's houses, shop-keepers houses, grosers, mercers, haberdashers, all manner of tradesmen that were inhabitants of that street lost both houses and goods by means of that consuming fire. Yet, that was not all, it burned downe the Market house and ran into St Marie's parish, and burned the church and many dwelling houses in that parish, so that in St Peter's parish and St Mairie's parish it is verified that at least three hundred families were dispossessed of their habitations, all which was done in the space of three or foure houres. For when the fire had fastened on one of the houses where were piles of wood and faggots in their backsides it flamed and burned so strongly that all that could ever be done could not quench the fire, until it had devoured and burnt to ashes, all these places I have here named.

Yet that is not all, for it was not the houses that were burnt alone, but also the goods that were in them; there was brasse and pewter, gold and silver melted, the value whereof cannot be made knowne, there was silks and tafety woollen and linen cloths and many other rich commodities consumed to ashes. There were foure or five tun of cheese which was laid in store in the market House consumed to nothing. It would make the heart drop tears of blood that had but heard the doleful cries and heavy moanes that passed between men and their wives, parents and children. The wife crying out to the husband 'Oh dear husband, what will become of us and our children?' the husband answering; 'We are all undone, I know not what to doe.'

The children crying for bread, the parents had none to give them, nor so much as a house to put their heads in, nor a bed to lay their weary limbs upon.

And thus were the poor made poorer, and some of the richest became as poore as the poorest. And now they are all in a sad condition, the Lord in his mercy send them comfort. Little did they that had plenty in the morning thinke that they should be made destitute and desolate before night...

This loss was so catastrophic that the Parliamentarian Council of State at Whitehall set up a committee of 30[2] to manage and order collections to be made in London, Westminster and the boroughs and principal places in England and Wales to distribute among those who had been sufferers in this great calamity. Because in excess of 250 buildings had been destroyed by the fire, clearly it would have been beyond the capability of the local masons and carpenters to repair the damage from their own, and local, resources. Tradesmen from far and wide would have been required, and it is thought that among these came the first Hancock, probably as mason or carpenter.

What is known of the generations[3] that followed suggests that the Hancocks were professional craftsmen who pursued a number of different trades. They were associated with clock making, coaching, and shoemaking among other occupations and, as their fortunes improved, they tended to marry the daughters of the better-off tradesmen or educated families. In this way, they gained a knowledge of the world that placed them quite well up the social ladder. However, it seems odd that no Hancock ever held civic office in the town throughout the whole of this period, and the only reference to a Hancock in the entire records of the Quarter Sessions[4] was to Giles Hancock, who was presented in 1743/4 for 'selling gin contrary to the Act, and keeping a disorderly house to the great fear and disturbance of many inhabitants.'

One hundred years after the fire, in 1753, James Hancock was born, and two bothers followed in 1755 and 1758. Smallpox swept through the town in 1759 and carried off James' father Philip and both his little brothers, leaving James as the sole survivor of that branch of the family.

In 1778, aged 25, and well established in his trade, he married Betty, the pretty daughter of Thomas Coleman, a maltster, whose family had been in that business for more than a century. The year 1780 saw the arrival of their first child, a daughter called Mary, after her grandmother Mary Brooker. Over the ensuing 20 years, 11 children followed, three more daughters and eight sons. The parish registers show that three of the daughters died before reaching puberty, all it is thought, from consumption.

By 1816 James Hancock, cabinet-maker, upholsterer, and timber merchant, was a man at the top of his trade, and master of a substantial business. His workshops were comprehensively equipped with excellent machinery, including three collar mandrel lathes, which between them could turn anything from the post of a four poster bed to the tiniest and most dainty spindle. On the walls were racks of hand tools, hallowed by age and use, many of which he had made himself over his lifetime. In his smithy was a forge and bellows[5].

In the 40 years he had been in business in Marlborough (Fig. 1.1)—from 1778 to 1818—he had witnessed many changes. Among these, perhaps the greatest had been the growth of traffic on the Great West Road. Marlborough had for centuries been an important staging post on the journey westward from London, catering for the horses and passengers of the lumbering old coaches that struggled along the rutted muddy tracks that in those days passed for roads. In more recent times, the rapid growth and popularity of Bath had brought about great improvements to the road surface and drainage. The reorganization of the mails by Ralph Allen and the introduction of the fast mail coach by John Palmer now meant that a letter posted in London could reach Bath the

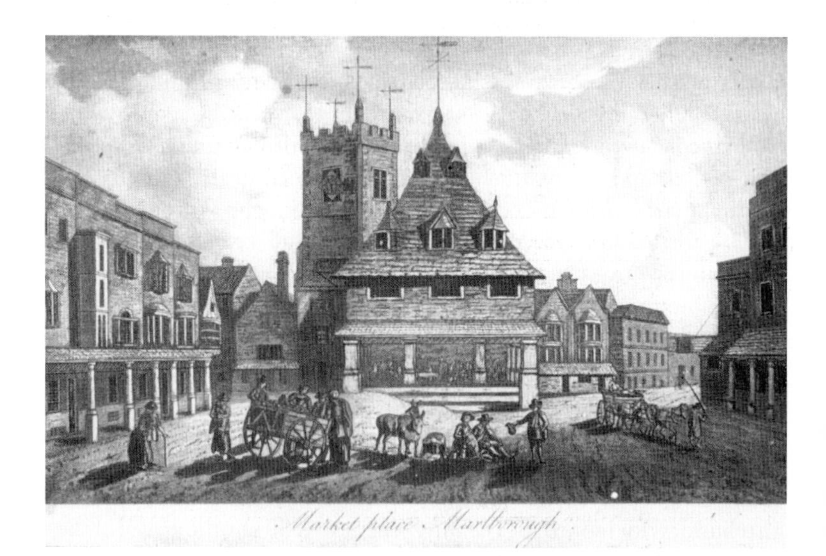

Fig. 1.1 Marlborough in 1800.

same day and a reply received back in London the following day. The growth of the fashionable carriage trade had driven upwards the standard of comfort and sophistication of the town's inns, of which the most famous was The Castle Inn, which could stand comparison with the finest in London or Bath. With improved communications came something else—news of the wider world—and news of new ideas.

In his lifetime, James Hancock had seen the demand for his furniture grow from simple country pieces of oak to the elegance of mahogany and even satinwood with the finest inlays of rare veneers. His workshops could provide everything necessary for gracious living, from carved panels to painted screens, furniture plain finished or japanned and even *trompe l'oeil.*

In these workshops his four elder sons, James (born 1783), Thomas (born 1786), John (born 1788) and William (born 1789) had, under his critical eye, all served their proper seven year apprenticeships in the craft of cabinet-making and upholstery (Fig. 1.2). He had been a

THE CHILDREN of JAMES HANCOCK & BETTY COLEMAN, married May 24th 1778 at Marlborough.

NAME.	BIRTH.	BAPTISM.	CHURCH ADMISSION.	DEATH.	BURIAL
MARY.	27th April, 1780	3rd Oct. 1780.		11th Nov. 1794	14th Nov. 1794. Marlborough.
JAMES.	27th May, 1783.	31st May, 1784.		15th Jan. 1859. Upper Homerton.	21st Jan 1859.
THOMAS.	8th May, 1786.	6th June, 1786.		26th March, 1865. Stoke Newington.	1st April, 1865. Kensal Green.
JOHN.	28th Feb. 1788.	27th March, 1788		13th March, 1835. Marazion, Cornwall.	Ludgvan, Cornwall
WILLIAM.	1st July, 1789.	15th July, 1789.		20th June, 1848. Milton, Gravesend.	
GEORGE.	30th June, 1791.	11th July, 1791.			
ELIZABETH	8th June, 1793.	17th Feb. 1795.		17th July, 1812.	22nd July, 1812 Marlborough.
MARY ANN.	22nd Jan. 1795.	17th Feb. 1795.		19th Oct. 1795.	16th Oct. 1795. Marlborough.
MARY	1st May, 1796.	6th Sept. 1796 (privately)	23rd Sept. 1797		13th March, 1800. Marlborough.
HENRY	27th Aug. 1797.	6th Sept. 1797.	23rd Sept. 1797		
WALTER.	18th June, 1799.	22nd June, 1799 (privately.)	28th Dec. 1802.	14th May, 1852. West Ham.	
CHARLES.	16th Dec. 1800.	26th Dec. 1800. (privately)	28th Dec. 1802.	30th July, 1877 Blackheath.	3rd Aug. 1877 Brompton.

Fig. 1.2 The children of James and Betty Hancock as recorded in 1924 by B. D. Porritt.

Fig. 1.3 Inlaid mahogany tray made by James Hancock.

stern master and would accept nothing less than perfection from them. When they finished their apprenticeships they were as capable as he of turning out the finest work (Fig. 1.3).

His eyes had not only overseen their training, they had also seen the newspapers which daily carried new ideas to Marlborough and which had opened his eyes to the rapid changes that were taking place on every side. He observed the beginning of the end of the old world of traditional crafts and the coming of the new; the growth of steam engines like the mighty pumps at Crofton on the Kennet and Avon canal nearby—engines which, from 1807, could lift a ton of water at a single stroke. There was news of a steam engine in the South Wales coalfields that could pull the coal drams along a railway, steam replacing the horse.[6] The name Trevithic was heard of and there were accounts of a demonstration of a road carriage powered by steam in London.[7] Everywhere the speed of change was accelerating, and James and Betty Hancock were far sighted enough to realize the importance of giving their children an education far wider than the one James himself had had.

In about 1980, there appeared at an antiques fair in Marlborough a beautiful tripod table with a reliable attribution to James Hancock. This was japanned black, with the top decorated in *trompe l'oeil*. It depicted a copy of Crabbe's poem of 1810, *The Village*, open at the following page:

> At her old house, her air the same,
> I see mine ancient letter loving Dame;
> 'Learning, my child' said she 'shall fame command;
> Learning is better worth than house or land,
> For houses perish, and lands are spent,
> In learning then excel, for that is most excellent.'

The question of where to send the boys for their education offered just two choices in the town. There was the Grammar School, an ancient foundation which offered a traditional classical education, or, from 1780 onwards, The Marlborough Academy. Here was something quite different. Founded by John Davis, a dissenting minister, it set out to provide an education for dissenters. It was to this academy that the Hancocks sent their children[8] and it not only gave them a first class education to university level but more importantly it taught them how to think for themselves—an ideal frame of mind for young men about to enter a fast-changing world.

Attendance at this academy opened their minds to other influences. While there is no evidence to suggest that James and Betty were anything other than loyal members of the St Peter's congregation, the growing potency of non-conformism influenced the parents to send their children to a non-conformist academy, a decision that was to have far-reaching consequences.

The Hancock Family Bible, which has recently come to light after almost two centuries of invisibility, and is now back in the care of the James family tells us much about another early influence that was to bear fruit abundantly in later years. It is very large, and the print is big enough to be easily read by candlelight. The pages are almost worn out by frequent use, many of them stained by drips of candle wax from the nightly readings, many of them heavily repaired and reinforced. To James senior, a strict adherence to the precepts of the bible was fundamental to acquiring spiritual perfection, and in cabinet-making of the first quality his training was aimed at acquiring perfection in the execution of any article to be made but the Bible also contains many clues to the lives of the family at this time. In the margins, from time to time, Walter (born 1799) and Charles (born 1800), the youngest sons, have signed their names in beautiful copperplate script. Walter had naughtily written some musical notes on the front flyleaf, and signed them, but the most poignant is surely a dried lime leaf, probably picked from the lime avenue on The Green. It reads simply:

Walter—July the 17th of 1812—Bessy died this same afternoon.

The leaf lies where Walter placed it over the following text from St Luke's gospel:

And thou child shall be called the prophet of the Highest, for thou shalt go before the face of the Lord to prepare his ways; to give knowledge of salvation unto his people, by the remission of their sins. Through the tender mercy of our God, whereby the dayspring from on high hath visited us to give light to them that sit in darkness, and in the shadow of death, to guide our ways into the way of peace.

Bessy was the only daughter to have survived beyond puberty, finally surrendering to the family curse of consumption in 1812, and her death, at the tender age of 19 (Fig. 1.4), ended her father's vain hope that she might have been the one to carry on his business after he had laid down his tools and entered into the long rest. She had entered upon her apprenticeship beside her brothers and would have completed it two years later had she lived. For many years after her death, a fine panelled mahogany door with beautiful inlays was treasured in the town as a memorial of her brief life.[9] It had been her apprentice piece, normally made at the end of her apprenticeship, but hastily yet perfectly completed earlier, as it became clear that her light was about to be snuffed out. Her elder brothers would have made her coffin themselves in their workshop, from the finest pieces of English elm from their father's stock. They would have finished it in the traditional country way, by smoothing the rough wood with a soft Savernake brick, which would have imparted to the elm a faint red glow.

This must have been a time of mixed emotions for James and Betty. On 13 May 1811 they had celebrated the marriage of James junior, their eldest son, to Elizabeth Lyne, but James forsook the family business and moved to Lombard Street, London, where his skills were soon in demand. In April 1812 their first grandchild, Elizabeth Lyne Hancock was born but, as with so many of the Hancock girls, she died in infancy, just 27 months old. A year later, Elizabeth gave birth to James Lyne Hancock, who was to have a considerable part to play in the story yet to come of the Hancock rubber business.

The family bible also gives more evidence of a mother's worries. Napoleon's shadow had been cast long and dark over England. The *Posse Comitatus* or Muster Roll recorded the names of every male between

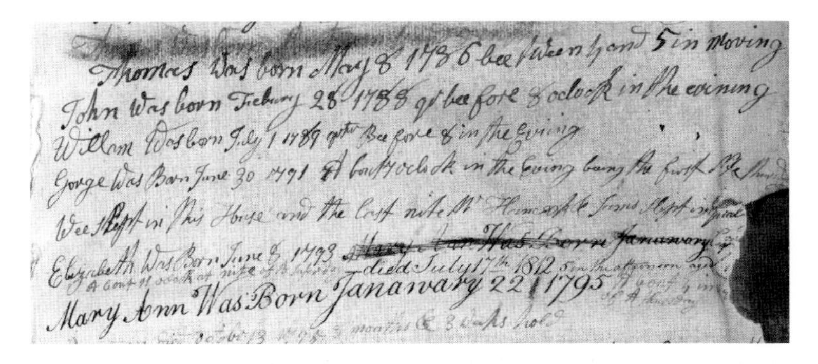
Thomas Was born May 8 1786
John Was born Febury 28 1788
Gorge Was Born June 30
Elizabeth Was Born June 8 1793 died July 17th 1812
Mary Ann Was Born Janavary 22 1795

Fig. 1.4 Details of inter alia Thomas' and Elizabeth's (Bessy's) births and the latter's death, from the inside cover of the Family Bible.

the ages of 16 and 60 who might be called up for military service. The Roll for Marlborough in 1803 shows James Hancock senior, James junior, and Thomas. On the rear endpaper of the bible there are various computations written where the ages of the other sons had been anxiously calculated in 1803 and again in 1805. James and Thomas were clearly liable and although we cannot now be certain, it seems likely from later events that Thomas certainly knew how to handle a rifle, and had knowledge of certain military matters, suggesting that he at least had undergone some basic military training. The colonel of the local militia then was the Marquis of Ailesbury, who also happened to be the freeholder of the properties that James Hancock rented in the town on the north side of the High Street, near the Town Hall, and this connection too was soon to prove significant.

The long years of the war with France had slowly dragged England to its knees, leaving the country much impoverished. Commerce was at a standstill, while food prices rose and fell erratically all over the country. The cruel frosts of 1814, which had brought hunger and distress to every poor man's door and had taken the price of coal to over £5 per ton, still haunted the memories of the labouring classes.[10] Now came something worse.

On 10 April 1815, far away on the other side of the world, Mount Tambora erupted. This was the largest eruption in recorded history, and its effect was to place up to 200 cubic miles of volcanic dust into the atmosphere. While the English were celebrating the victory over Napoleon at Waterloo, the effects of Tambora were gradually spreading round the world; a hidden menace which nobody then perceived.

In the year 1816, many things conspired to reduce the value of labour. Manufacturing operations were paralysed, agricultural produce was suddenly reduced to half its former value; and farming operations were altogether suspended in many parts of the kingdom. Labour everywhere failed to obtain remunerative employment. The navy had suddenly been reduced from 100,000 to 33,000 men, the militia had been disbanded, and the establishment of the regular army much contracted—at a time when the supply of workmen was already in excess, some 200,000 able bodied men had been added to the crowd seeking employment. The hopes which the peace had raised had been suddenly turned into the most bitter disappointment. Distress was visible in every class of life. The home trade was at a standstill, landlords received no rents, their tenants could sell no corn. Thousands of acres of agricultural land were offered rent free to anyone who would take them, but none accepted them. Distress pervaded every branch of commerce, nothing seemed wanting to complete the universal misery and distress; but the cup of suffering was not yet full. If labour was scarce, bread had all at once become cheap, and in the cheapness in the chief article of food, the labouring community found some moderate consolation.[11]

By now though, the effects of Tambora were becoming all too apparent in Europe. An icy wet spring in 1816 threatened the corn harvest, thus depriving the poor labourers of their brief consolation.

The price of wheat rose to eighty shillings a quarter, and the old prices seemed on the point of returning. Farms which had been thrown out of cultivation were re-let; the incoming tenants mistaking the rise of prices which was the threatening of dearth, for the effect of war creating an excessive demand. Their hopes were speedily dashed.[12]

'The rain it raineth every day' or '1800 and froze to death' was the universal refrains in the summer of 1816, soon to be known as 'the year without summer'.[13]

The harvest was one of great deficiency, and the two following years the crops did not exceed the average; high prices of provisions were the consequence which, combined with the low wages and scarcity of employment, produced the greatest discontent among the working classes.[14]

Bread riots followed which had to be quelled by the military. It was not a happy time. With the death of Bessy, coupled with the gloomy economic and social climate, any hope of the family business continuing seemed now to evaporate. James had been in London for several years and the three next older brothers, Thomas, John, and William, had followed to seek their fortunes, Henry and George, the middle brothers, were given their inheritance and left, according to family tradition, for

America while Walter had begun his apprenticeship to Cutmore, the famous London clockmaker. The choice of Cutmore's was no accident. It was the Protestant work ethic—'time must not be wasted'—which initially made the Calvinist jewellers of Geneva turn from jewellery towards clock-making and, as their workshops spread to England, so did their philosophy which was so close to Thomas' heart.

Charles was later to follow his brothers to London in 1817 to be taught by James Stark, then at the Royal Academy schools. Stark had been a pupil of John Crome of the Norwich School of painters,[15] the foremost influence in painting in England. The days of Hancock and Sons in Marlborough were drawing to a close.

On the 27 and 28 April 1818, 165 years after the Great Fire of Marlborough to the very day, the remaining stock in trade of James Hancock was sold. It comprised

very neat and elegant mahogany chairs of the best materials and superior workmanship, japanned cottage and chamber ditto, commodes, chests of drawers, bedsteads, wash stands japanned plain and ditto, dressing, night, and claw tables, easy chairs, counting house desks, bookcases and bookshelves, handsome four post bedstead, rich carved mahogany pillars, field ditto with bedsteads of other descriptions, capital collarmandrel lathe and two small ditto, three frame pit saws, two long ditto no frames, crosscuts, bottle and wedges, grindstone, forge and bellows etc. etc.[16]

A footnote added:

James Hancock impressed with the grateful remembrance of past favours, returns the public his most sincere thanks for the liberal support he has experienced over the last forty years.

James and Betty, having now no one to look after but themselves, moved from the High Street to a small house in the Back Marsh. James did not live to enjoy a long retirement or to see the achievements of his children that were to transform the world. He died, aged 68, in 1821 and Betty followed him in 1836.

With the cessation of the business, the doors close on the Hancocks in Marlborough, and open again in London, where the sons of James and Betty are now gathered, equipped only with their education, their faith, and their trades to seek their fortunes in post-Napoleonic England.

2

The Hancocks Gather in London

Thomas finds his spiritual home—William Huntington—his early life and rebirth as a preacher—effect on Thomas. Walter begins his apprenticeship as a clockmaker but William moves to Bury St Edmunds as a cabinet-maker, etc.—becomes bankrupt. Thomas' and John's first business and patent—the Pilentum or Ladies Accelerator—Thomas discovers rubber—early rubber history—Cortez, la Condamine, Fresneau. Joseph Priestly coins the name 'Indian Rubber'—Thomas invents his pickle or masticator to convert scrap rubber to a useable material—sets up his factory in Goswell Mews—decides to keep pickle secret rather than patent it—involvement with Michael Faraday. Walter and Charles join his business venture.

When the eldest brother James left for London in about 1812 it was not surprising that he should still feel a spiritual affinity with Marlborough, where all his family still lived. When his first child died in 1814, he accompanied the little coffin on the coach back to the town so that she should be buried in the shadow of St Peter's tower, next to his sisters and surrounded by his ancestors.

Thomas seems to have carried no such emotional baggage, and struck out on his own from the start. Throughout his long life he followed a recognizable pattern of behaviour which was to prove singularly effective. He had an exceptionally clear mind, as events will show, but when in need of professional advice to deal with a problem that was outside his own experience, he always chose the best, regardless of the cost. He had grasped the universal truth that good advice is always a good investment.

The same principles led him to seek out a new spiritual home among the confusing choices open to him and his brothers in the teeming metropolis. The daily readings from the Bible at home had prepared the ground to be further fertilized by his studies at the Marlborough

Academy, where non-conformism was the order of the day. He was ready for a crop to be sown, but where was the sower?

Whether we believe in coincidence, synchronicity, or the mysterious hand of Providence, we now encounter a most extraordinary chain of events that was to guide Thomas throughout his life and exercise a most profound influence on him and those who knew him for long after. These events also played a very significant part in his commercial success. Before this episode begins we have to start at the end of Thomas' life with part of an oration at the Langham Chapel later privately published in 1865 under the title *A Discourse Occasioned by the Death of the Late Thomas Hancock.*[1] The oration was given by William Benson, a young architect and non-conformist preacher, to whom Thomas had been very well known. 'What factors', he enquired rhetorically, 'influenced Thomas' character?' 'Surely' he continued 'that which he heard when young from the mouth of that great and blessed servant of Christ, William Huntington, was like good seed sown in good ground, for it sunk deep into his heart and influenced all his life thereafter'.

There are many people today to whom the name of William Huntington[2] is completely unknown, but in London in early 1813, indeed anywhere in the south of England, it would have been difficult to find anyone to whom the name Huntington was not known and he was destined to become the sower that Thomas Hancock was seeking.

Huntington styled himself 'Celebrated coal heaver and sinner saved'. His childhood had 'been embittered by sufferings of hunger, cold, and nakedness to that extent... he often wished he had been born a brute, so that he might have had sufficient to satisfy the cravings of nature'. From such miserable beginnings, his life had spiralled downwards with a certain grim inevitability. He had been forced to change his name from Hunt to Huntington to escape discovery by the family of a servant girl he had got into trouble. Having committed perhaps more than his fair share of sins and having only the bible for company, his feelings of guilt led him to seek some spiritual comfort. He tried attending a number of churches without finding any solace until he came under the influence of some Calvinists in Kingston upon Thames. In his twenty-fifth year, after a succession of menial jobs including heaving coal on the wharves of the Thames, he claimed to have married a Dorset girl (though no record of this has ever been discovered) but his misfortunes followed him. His first child died in infancy.

This death brought him to a serious reflection upon his life. Deep sorrow for sin wrought in him a determination to forsake once and for all his sinful companions among whom he had been the life and leader by his ready wit and jocular conversation.

In 1773 he had experienced what he perceived to be a powerful divine revelation, which had convinced him that he must bear witness to the truth of what he had received and that, from then onwards, his life must be devoted to preaching to any who would hear him. His lack of education, combined with a detailed acquaintance with the language of the Bible, gave his words a strange flavour of power and authority, which he developed into a highly effective mode of address. To this he added a hypnotic habit of slowly moving a capacious white handkerchief from side to side before him to emphasize his words from the pulpit. He gradually became a charismatic preaching phenomenon and thousands drew near, wishing to hear him. In the year of Thomas Hancock's birth, he had preached in Bath to such effect that his visit was talked of for a generation afterwards.

> Queues began to form more than an hour before he was due to appear. They hung on to windows inside and outside and got anywhere, just to hear the sound of his voice This visit was much blest to many, and is still referred to by some whose ancestors speak of it as an extraordinary revival.

In one of his letters to a friend, Thomas Hardy[3] wrote that 'The best Christians I generally meet with are Huntingtonians', and the fact that he wrote this nearly a century after Huntington's death bears strong witness to the continuing influence his preaching had on the hearts and minds of those who followed him. J. C. Philpot,[4] who quoted the observation by Thomas Hardy, offers a further description of Huntington's followers:

> There is, or as we must now say was, for so few of them are left, a depth and clearness of experience, a savour and a sweetness, a rich tender feeling, an unctuous utterance, a discrimination between law and gospel, letter and spirit, form and power, a separation from a lifeless profession, whether presumptuous or pharisaical, which distinguished them in a most marked and decisive manner, as a peculiar and separate people.

Huntington's fame would not have escaped the proprietors of the Marlborough Academy and thus his name, if not his preaching, would have been familiar to the Hancock brothers and so it is not surprising that they joined his congregation at his Providence Chapel, which was then in Gray's Inn Lane. Benson in his funeral discourse recalled that the first words Thomas had ever heard from Huntington were 'The Lord recompense thy work, and a full reward shall be given thee...' and seeing the great man arrive at his chapel in a coach drawn by four horses, it must have seemed to the Hancocks that here was a simple man, a deeply repentant and self-confessed sinner, whom the Lord had indeed recompensed and to whom, after a life of tribulation, a full reward had indeed been given. William Huntington did not live to see the close of 1813, but among his congregation were several people whose lives were to cross and recross the paths of the Hancocks and whom we shall meet again, especially a little Frenchman named Joseph Francis Burrell.

By 1815, all the brothers except Charles were now in London. James, Thomas, John, and William were earning their own livings, while Walter was in the first year of his clock-making apprenticeship where he would receive the only theoretical training available at that time to a prospective engineer. William was not able to settle down in the metropolis and very quickly moved on to set up in business in Bury St Edmunds. A notice in the *Bury and Norwich Post* of 20 December 1815 announced:

> Hancock, cabinet maker, upholsterer, auctioneer and undertaker begs respectfully to inform the nobility gentry and public of Bury and its vicinity that he has just commenced in these lines. And from his extensive experience in London and its neighbourhood feels confident of giving every satisfaction to those who may favour him with their support.
>
> N.B. Paper hangings, decorations, colouring in distemper, verandas, temporary rooms, summer retreats etc, every description of furniture in the cabinet making and upholstery business.

By 1816 he was advertising in the same paper that he had just received a genteel and fashionable assortment of paper hangings and carpets which he begs to offer on the most reasonable terms. A footnote adds that an apprentice is required to whose morals the strictest attention will be paid.

He was soon to discover a great truth. The quickest way to bankruptcy in those days was to supply goods to the nobility, who were not usually over-anxious to settle their bills during their lifetime. In 1820 he was to suffer his first (but not his last) bankruptcy, two years later receiving his discharge certificate having paid a mere three shillings in the pound.

While William was content to follow the traditional craft of his father, it is clear that Thomas and John were not afraid to experiment in new fields. They first appear in business together in 1816 as 'Hancock and Co., Patent whalebone Coach-makers in 55 St. James' Street'[5] (which runs between Piccadilly and Pall Mall). Later, in 1817, Hancock & Co. is listed at two different addresses, 8 Little Pulteney Street, where they are described as whalebone merchants, and 55 St James' Street, where they are listed as whalebone carriage makers. By 1820 three directories list them in St James' Street, but Robson's Directory[6] interestingly describes them as 'elastic whalebone manufacturers' and gives a separate address for John Hancock & Co., Coach Makers at 1 Little Pulteney Street. In view of coming events it is important to note the word 'elastic'.

In 1813 John Hancock had taken out a patent (EP 3733) for the use of whalebone in the construction of coaches, gigs, and carriages. Whalebone was not then a term used to describe the bones of whales but a substance derived from the mouths of Right whales—similar in strength to the hide of rhinoceros—and its commonest use was for the stays in ladies' corsets. From the patent it appears that whalebone was proposed for the spokes of the wheels and to strengthen and brace the carriage. For a period this material enjoyed a brief popularity in the coach- and sledge-building trade.

Interest in coach-building had received a great boost from the exhibition in 1816 of Napoleon's dormeuse at the Egyptian Hall.[7] After its capture by Major von Keller on the 18 June 1815 it was sold to a William Bullock, who brought it to London. Because Napoleon had been the bogeyman of Europe for so many years but was now safely in custody, the English public felt brave enough, albeit with a shudder of apprehension, to approach this mighty vehicle. Finished in dark blue with a frieze ornament of gold and with the imperial arms emblazoned on the bullet proof door panels it was a magnificent construction. Equally convertible inside to bedroom, field kitchen, office, bathroom, dressing room, or dining room, the crowds who pressed in the Hall to inspect it could easily imagine the great man himself sitting in it. They could draw out the doors of his desk and handle the pens and inkstand that he had so recently used, inspect and play with the many secret drawers which contained telescopes and maps and thrill at the holsters on the inside of one of the doors which contained a pair of pistols and in another holster adjacent to the seat they would find a further double-barrelled pistol. The frisson of terror felt by the visitors sustained such interest in this exhibit that when the doors finally closed even William Bullock must have been surprised at the £35,000 that had been subscribed. The carriage was acquired by Madame Tussaud's in 1842 and it remained in the collection for some 80 years. On 18 March 1925 a disastrous fire swept through the museum, destroying not only the carriage but also its setting: the Napoleon Room. Much else was also consumed by the flames. In 1976 the pathetic remains of this once splendid vehicle—a single heat-warped axle—was presented to the Museum at Malmaison.

In this perfervid climate, Hancock & Co., in about 1818, took advantage of the rising interest by offering, in addition to more conventional products, a most original vehicle, almost certainly designed by John Hancock.

Called the 'Pilentum' (a Roman horse-drawn carriage for ladies of a superior class) or 'Ladies' Accelerator' (Fig. 2.1), it was a tricycle specially designed for persons of rank and fashion. It was powered by two treadles acting on a crank fixed to the front wheel (a principle they knew well from the lathes in their workshops) and the power from these could be augmented by the additional use of hand cables. As the promotional literature[8] stated:

> This elegant little vehicle is peculiarly adapted for the use of ladies as well as gentlemen. It is impelled by the slightest touch of either the hands or feet, at a rate truly astonishing, and is so completely secured from upsetting that the most timid person might use it with greatest confidence.

It was certainly an elegant improvement on the only other form of personal wheeled transport then available, Denis Johnson's Hobbyhorse,

Fig. 2.1 The Hancock Pilentum or Ladies' Accelerator.

which had appeared in 1819 and was most certainly not suitable for young ladies.

Although there are no known existing examples of this curious and original vehicle it must either have made a strong impression on the public mind or the Hancocks were extremely adept at manipulating the advertising opportunities open to them for no fewer than three separate engravings of the Pilentum appeared. In addition to these a commemorative mug bore the transfer of a lady riding the Pilentum through the countryside and a plate was also available bearing a different image of this vehicle over a motto which read: 'Wonders will never cease'. Finally, Robert Cruikshank drew, and G. Humphrey published, a cartoon entitled 'The hobby horse Fair', which raises the intriguing possibility that Hancock & Co. also built an enclosed version which would have been somewhat similar to a sedan chair on three wheels. Perhaps it was no coincidence that Humphrey was also resident in St James' Street.

Thomas Hancock now had other things on his mind. In 1856 he wrote his *Personal Narrative* in which he described his life's work in the rubber industry:

> I have no very clear recollection when I first began to notice the peculiar qualities of India rubber, but well remember that the more I thought about it and tested its properties, the more I became surprised that a substance possessing such peculiar qualities should have remained so long neglected, and that the only use of it should be that of rubbing out pencil marks.

In truth it had been used for rather more that that, but by making the statement Thomas was confessing complete ignorance of what had gone before. The earliest certain reference to rubber, or *caoutchouc* to use the Mayan word by which it was universally known until the late eighteenth century, was by Hernando Cortez in 1519 who, with his conquistadors, saw numerous products made of rubber in the court of Montezuma, including vessels made by dipping clay formers into latex and smoking the resultant skin to sterilize it, in the same way that we cure fish or ham today. Rubber torches were burnt for light and to keep evil spirits at bay, and his soldiers were taught by the Aztecs how to paint their capes with latex to make them waterproof. He also noted that of the various taxes levied by Montezuma on his subjects, one was for 1,600 balls of rubber per year. More recent archaeological studies have shown that rubber had been used in Mesoamerica since at least the seventeenth century BC.[9]

The arrival of the conquistadors brought an end to the Aztec civilization and their rubber technology, leaving only the primitive natives in settlements along the Amazon and its tributaries to preserve some practical applications.

Two hundred years after Cortez, a young Frenchman, Charles Marie de la Condamine, found himself in Central America as part of a French expedition to take scientific measurements to determine the size of the Earth. Returning to Cayenne via the River Amazon, he found natives using a variety of rubber articles and sent a selection to France in the hope that the king might be interested. He was not. In the meantime he met another Frenchman, Francois Fresneau, who became fascinated by this new material. La Condamine returned to France and Fresneau set out to investigate its origins and uses, eventually writing a paper which la Condamine presented to the Royal Academy of Science in Paris in 1751. This was the first scientific paper on aspects of rubber and could be considered the start of European interest in the material. For that paper Fresneau was to become known as 'the father of rubber'.[10]

A major problem with the new wonder material rapidly came to light. It was virtually impossible to ship the liquid latex from the tree to Europe because it was too unstable and curdled like stale milk. When Fresneau eventually returned to France he continued his research into ways of dealing with this problem, and in 1761 he discovered that pure oil of turpentine was the ideal solvent with which to prepare rubber solutions.[11] These could then be used in Europe, like latex in Amazonia, for the manufacture of dipped or coated articles.

It was soon after this that rubber received its British name. It had previously been referred to as *caoutchouc*, a phonetic interpretation of the Mayan words generally interpreted as 'weeping wood' but in 1770 the famous scientist Joseph Priestly saw a ½-inch cube of a new material in Mr Nairn's artists' materials shop and asked what it was. On being told that it was excellent at removing unwanted charcoal or pencil lines from sketches by rubbing over them and that it came from the Indies he called it an 'Indian rubber'.[12]

Over the next 20 years rubber solutions were used widely for medical devices, to coat the fabric of hot air balloons and also the first hydrogen balloon. In 1790 the first patent relating to rubber was issued to Roberts and Dight.[13] This was for the use of a solution of rubber to prime artists' canvasses before applying oil paint. Stretched and loose canvasses which had been so treated remained in catalogues such as those of Winsor and Newton until late in the nineteenth century.

It is ironic that as a youth, Thomas' brother Charles could have been painting on these canvasses before Thomas became aware of the material with which he was to change the world.

Continuing with his *Narrative*:

I had spent my earlier years in mechanical pursuits, and was well acquainted with the material generally employed therein, and also with the use of tools, so far as to enable me to make with my own hands almost any kind of machinery required to carry out my views; but of chemical knowledge I had none. I premise this because it will be seen in the course of my narrative, that although the substance I was contemplating apparently required to be treated chemically, I owe my success principally to the practical knowledge and degree of skill I had acquired in mechanical manipulation, which proved eventually to be the best preparation I could have had for operating upon rubber; and it is a singular fact that, although this substance had attracted the notice of chemists from the earliest date of its importation into Europe, and some of the ablest had employed themselves upon it, they failed to discover any means of manufacturing it into solid masses, or to facilitate its solution.

As well as missing the work of Fresneau in France, and Fabrioni in Italy a few years later, he also appears to have had no knowledge of the experiments of James Syme in Edinburgh in 1818 and Charles Macintosh's earliest work in 1819. However, some time in that latter year, Thomas came across a patent (No. 3718) which John Clark had taken out in 1813 for 'beds, pillows cushions etc.' in which Clark described how fabrics could be made airtight by treating them with a solution made by immersing rubber in spirits of turpentine, and then boiling the mix for several days in a large excess of linseed oil.

Patent law was then, as it is today, a dangerous jungle in which prowled patentees and their lawyers. With his customary caution, in view of his proposed experiments with rubber, he consulted Mr Bolland, (later to become a Baron in Chancery), who was then the greatest living expert on patent law and who was to advise him in this sphere for the rest of his life. It was suggested that he should try to buy the benefits of this patent if he could, and later in the same year he succeeded in acquiring them for £10.[14] Clark of course had already discovered what Thomas was about to; that the process was useless, but the purchase had one great advantage—it enabled Thomas to continue with his experiments without fear of retribution.

Thomas recalled:

I was at first imbued with the notion that to make it [rubber] useful, I must first find a good solvent; and I think my first experiments were directed to some attempts to dissolve it in oil of turpentine, but I found that I could make only very thin solutions and these dried so badly, or rather not at all, that they were useless. The oil of turpentine then procurable was no doubt of inferior quality; when pure it dries perfectly. This was in 1819.

The following year, from Stoke Newington, he applied for his first patent (EP 4451) for 'an improvement in the application of a certain material to various articles of dress and other articles, that the same may be rendered more elastic'. This patent involved the incorporation of rubber rings, which he cut from rubber bottles, which were being imported from Brazil, into articles of clothing to make them elastic. Given that he was a coach-builder he would have heard of, or experienced first hand, the miseries to which outside coach passengers were subjected on long journeys. Rubber rings set into the cuffs of coats would have gone a long way towards stopping water running up the sleeves. His first idea of simply sewing the rings into the fabric was soon abandoned when he realised that the rubber tore at the needle holes. He tried increasing the thickness of the rubber where the stresses were greatest but this also proved ineffectual. He then attempted to achieve his purpose by rolling the cloth round the spring and then sewing up the tube of cloth with the springs inside so that needle holes in the rubber were avoided. The failure rate was still unacceptable. A lesser man might at this point have given up but his continuing interest in the peculiar properties of rubber drove him on. He had noticed that by constantly expanding and releasing the springs, as he called them, they suffered from a form of fatigue; the fine cuts left in the surface as the rings were shaped were growing until they eventually snapped.

This appeared a very formidable obstacle...but was soon overcome, for I observed that some of the springs were, and some were not, affected in this way; and on tracing back the steps which had been taken with the two kinds, I found that those springs on which I had used boiling water after they were cut, did not crack at the edges...

This elegant piece of observation reveals the advanced scientific methods he was using well ahead of his time, as he was readily able to trace the life of every spring he had made as well as the individual treatment it had received. Later in his career this accurate recording was to save him in the greatest of the many legal challenges with which he was confronted.

By now his workshop was piling up with small pieces of rubber for which he had no use. As orders began to increase for his elastic cuffs so did the growing mountain of scrap rubber. Looking at these wasteful piles of spaghetti-like threads and meditating on how he might be able to operate on them, an idea surfaced from his rural youth in Marlborough, where seemingly impossibly tangled sheep fleeces could be tamed by the use of carding brushes whose spikes could tear through the tangled fibres reducing them to useable order. He may even have encountered one of the new carding machines that consisted of a drum covered in spikes rotating in a spike-lined box, thus combing the fleece by a rotary motion. He had already discovered that if a piece of rubber was cleanly cut and then the cut faces immediately reunited they would join together, but if they were left separated for a short time then the mutual adhesion vanished. Could such a machine be made to tear the scraps of rubber apart so that the continually generated fresh surfaces would fuse together? Now his ability to make any mechanical device to carry out his designs came into its own and he set about constructing his prototype.

So was born the 'pickle' (Fig. 2.2), which was later to be called the masticator. Although Thomas did not know it at the time, and indeed

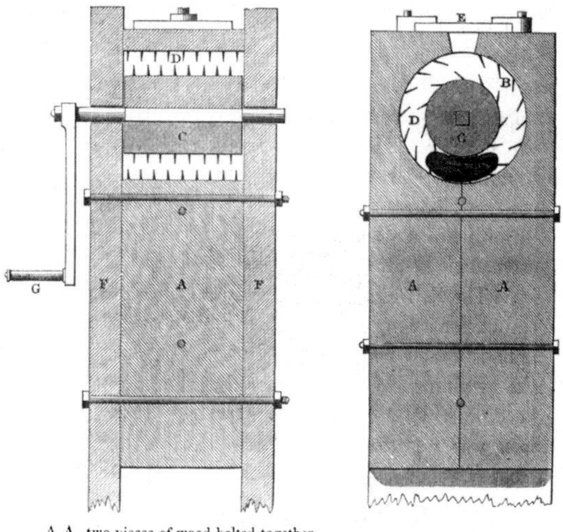

A A, two pieces of wood bolted together.
B, a hollow cylinder cut out of A A, and studded with teeth.
C, a cylinder of wood studded with teeth, and having a spindle passed through it.
D, space between the two cylinders B and C.
E, an opening with a cover.
F F, two pieces of wood bolted on both sides of A A, and enclosing the space D, and cylinder C.
G, a winch.
The darkened spot in space D represents the charge of rubber.

Fig. 2.2 Thomas Hancock's prototype 'pickle' or masticator.

it was to be a century before the proposition was advanced by Herman Staudinger, natural rubber is a long-chain molecule called a polymer which is built up of many thousands of small units joined together. It can most simply be visualized as the children's favourite—a 'poppet' necklace. What Thomas' pickle was doing was masticating or 'chewing up' these long polymer chains, which enabled them to untangle themselves more easily from the individual scraps of rubber and combine into one large piece. If one imagines that an open 'poppet' necklace represents the long-chain polymer of natural rubber as it is taken from the tree, then that chain could be built up of 100,000 small units or more. Masticating it so that lengths are typically in the 2,000–20,000 unit region gives a much more amenable material without doing significant damage to its elasticity. Although the crude rubber that Thomas put into his 'pickle' was initially tough, and the effort of turning the handle very great, he soon felt resistance building up and heat generated by the friction. Looking anxiously inside this is what he found:

> ...a ball of rubber. This ball when cut open presented a marbled or grained appearance; the union of the pieces was complete; the graining exhibited the pieces curiously joined together...the whole had become a homogeneous mass.

This was the eureka moment for the future of the rubber industry.

> I now had at my command the means of reducing all kinds of rubber of whatever size or form the original pieces might be, to a solid mass, without any foreign admixture or the use of any solvent....The discovery of this process was unquestionably the origin and commencement of the India rubber manufacture, properly so called; nothing that had been done before had amounted to manufacture of this substance, but consisted merely of experimental attempts to dissolve it, and even this had never yet been effected for any useful purposes.

Immediately a larger version of the masticator was ordered from Messrs Topham and Hague.[15] Further experiments led him to place the hot rubber from the mill straight into cast iron moulds by which he could, by the application of pressure, then make uniform blocks of rubber. He soon found that these could be cut by a water-cooled knife to make slices of rubber, some of which were thin enough to be translucent.

He was now so completely confident that a great future in rubber manufacture lay before him that in 1820 he took premises in Goswell Mews, off Goswell Road, where the company was to remain for over a

100 years. Larger machines were ordered, including roller mills, while power was soon supplied by a four-horse engine, but now came another problem. Should he patent his masticator? The cost of a full patent for England, Wales, and Scotland was about £400. This was a small fortune, which he did not possess and, anyway, patenting the masticator would let his secret out the bag. It was an agonising decision and if the wrong choice were to be made, he might lose the entire advantage that he perceived he had gained.

Here he calculated that he had one unlikely advantage that was not open to any potential rivals: his membership of the Eschol Chapel, successor to William Huntington's Providence Chapel, which was now under the guidance of Joseph Burrell. To the members of this closely knit community who shared his Calvinistic beliefs he felt confident of entrusting his priceless secret and in this he was completely vindicated, for the workmen he recruited from Burrell's congregation kept it safe for the next 13 years.[16]

After explaining the need for complete secrecy, Thomas instructed them to refer to his machine as 'the pickle' and the process as 'pickling' to throw any enquirers off the scent and to mislead them into believing he must be using some sort of solvent.

In 1824 he supplied some sheets of rubber to Michael Faraday, who carried out an extensive analysis of them, and in 1826 reported his findings in the *Quarterly Journal of Science and the Arts*. Faraday, some 15 years younger than Thomas, was to be a life-long friend although it is not known whether they first became acquainted through their religious beliefs or their membership of scientific organizations! Faraday was a devout member of the Sandemanian Church, a fundamentalist Christian order that demanded total faith and commitment and no doubt his dissenting views sat well alongside those of Thomas.

By now Thomas' business was expanding rapidly. As his range of new products expanded he was able to recall:

> Drawing masters and others using black lead pencils approved highly of the neat square blocks cut from masticated rubber, and the quantities in demand for this purpose constantly increased; they are still supplied to a large extent. Pieces, blocks and forms of much larger dimensions began to be enquired for, made for purposes with which I was not acquainted. The edges of wheels, and surfaces for rollers, and cylinders were also covered with rubber of various thicknesses for machinists. The billiard table makers also applied for long evenly cut pieces to form the cushions of their tables, which were successfully applied and have continued to be used for that purpose (with modifications)

ever since, to the exclusion of all others. In the early part of 1822 I began to make tubing of the sheet rubber, and afterwards with alternate plies of cloth coated with solution; some were also covered with leather, velvet etc...

The speed at which the business was now growing meant that Thomas could no longer cope on his own with the many and varied new demands he was receiving, particularly since they often required new designs of machinery. Walter had by this time just completed his own apprenticeship and although he had just set himself up as a jeweller in King Street, off Northampton Square,[17] he was easily persuaded that a brighter future lay in assisting Thomas with the design and assembly of these new machines. Shortages of capital must have posed a serious problem for them and it might have tempted another man to take in partners but Thomas was instinctively against this and so he invited his youngest brother Charles to join him. It kept the business in the family—or did it?

3

From Seawater to Steam

The innocence of Walter and Charles in the ways of business—new partners and disaster. Charles returns to painting, Walter to engineering—invents his 'pistonless' steam engine and becomes hooked on steam—builds a series of ever-more ambitious steam carriages and becomes the first person to ply for trade with a steam carriage in London—problems and tribulations—closure of his business and birth of his daughter.

The arrangements ran smoothly enough to begin with but in 1823 Thomas patented a process for producing a rubberized anti-fouling paint for waterproofing the bottoms of ships and protecting them from the ravages of marine worms, a constant and serious problem in tropical waters. Because he had no time to develop it, he gave it to Walter and Charles. In the early stages, the process promised good commercial returns but that very promise was to bring about its downfall.

Here it needs to be remembered that Walter was only 24, and Charles 22. Of commerce they knew nothing and not much more about human nature, especially where money was involved. Neither had undergone the rigorous apprenticeship to their father nor had they had the opportunity of attending a full course at the Marlborough Academy. It is also possible that Charles, being the youngest and last to leave home, had been somewhat over-indulged. Whatever the cause, Charles was certainly different; having set out upon an apprenticeship to fine art where a painting took a long time to finish and the returns were at best speculative for a young and unknown artist he now found himself living in a household where the conversation positively fizzed with talk of possible applications of rubber to innumerable inventions. This atmosphere distracted him from the single-minded pursuit of excellence while the attractions of industrial processes, which seemed to him to offer unlimited opportunities for acquiring an easy fortune, constantly tugged at his sleeve. He had exhibited a number of paintings including some portraits at the Norwich Society between 1818 and 1821,[1] which drew from the critics the observation that he displayed an early and rising talent. The Royal Academy list of exhibitors shows that in 1819 he had managed to get a portrait of his brother John hung at the Summer Exhibition, and during this period he undertook a series of portraits of members of his family, any of them that is who would consent to sit still for long enough, which revealed a talent that steadily developed from the naïf to the accomplished.

Because Walter and Charles lacked experience in industrial concerns they were easily persuaded to allow partners to bring capital into their new and profitable process. Nobody had apparently taught them the old proverb, that 'he who pays the piper calls the tune'. The partners saw at once that the two lacked any business acumen and disputes arose almost at once. These were found to be irreconcilable and litigation followed,[2] which lost the Hancocks the process and the advantages they thought they had secured. Thomas meanwhile watched these proceedings, but kept his own counsel.

This was a major setback for Charles and Walter as they had been putting all their energies into this part of the business and there was little to fall back on. The business at Goswell Mews continued to thrive but Thomas felt disinclined to pass any more new projects to the Stratford factory. Charles was obliged to return to his painting and, in the belief that portraits were where the money was, returned to the improvement of his techniques. He did, however, find time to paint his mother, Betty (Fig. 3.1).

Fig. 3.1 Betty Hancock painted by her son Charles.

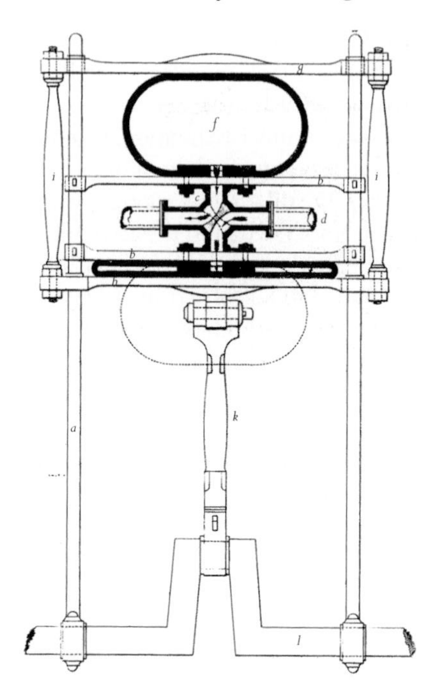

Fig. 3.2 Walter's pistonless steam engine.

Walter returned to his workshops and cast around for a new challenge. One area in which he was interested was increasing the rate of manufacture of the valves for Thomas' various rubberized fabric inflatable products but for this he needed to replace his man-powered lathes with machine-powered ones. As no suitable small engine was available, he set his mind to devising one and came up with something quite remarkable and completely original.[3]

What triggered his design is unknown but he must have been thinking of one of Thomas' inflatable lifejackets because the basis of the engine was a pair of rubberized cloth balloons or lungs. In Fig. 3.2 the top one is shown inflated and the lower one deflated. Steam is supplied through a two-way cock to the two lungs alternately and the valve is designed so that when one is filling, the other is venting to the atmosphere and can thus be compressed. This results in a reciprocating movement of the supports of the bags, which is translated to a rotating of the drive shaft *via* the crank. In 1838 Walter wrote:

...an engine of this construction, of four horsepower, has been employed at the author's manufactory at Stratford for some time and worked most

Fig. 3.3 Isaac Newton's steam concept car.

satisfactorily, its simplicity, comparative cheapness and diminished friction are its principal recommendations.

The immediate success of his lightweight engine gave Walter pause for thought. Would such an engine be able to propel a carriage?

There had been various efforts at making a vehicle propelled by its own steam engine[4] although the *'concept jet car'* designed by Isaac Newton (Fig. 3.3) never got off the drawing board—possibly for health and safety reasons! That did not stop a cartoonist illustrating the sort of life a ploughman could look forward to if it ever came to fruition (Fig. 3.4)! In 1769 the French engineer and mechanic, Nicolas Joseph Cugnot, invented and built a military tractor which had a top speed of 2½ mph and had to stop every 10–15 minutes to build up steam pressure. However, the steam engine and boiler were separate from the rest of the vehicle and placed in the front, so it hardly counted as a self-contained steam-powered vehicle. Nevertheless, the following year he built a steam-powered tricycle that carried four passengers. Incidentally, within a few months he had driven one of his steam vehicles into a stone wall, making him the first person to be involved a motor vehicle accident!

Walter wrote that he was aware of the steam carriage built for Julius Griffiths of Brompton Crescent, which he had patented in 1821 and which he had built by Joseph Bramah, the celebrated engineer and manufacturer. This carriage failed from the perennial problem of an inadequate supply of steam. Walter may also have been spurred on by reading about Burstall and Hill's remarkable steam carriage, built in 1824, which was so heavy it shook itself to bits on the existing common roads. Coach-building was an art well understood by the Hancocks,

Fig. 3.4 Steam ploughing in a perfect world.

so perhaps it was natural that Walter should fall back from his sad adventure into industrial territory to more familiar and safer areas. Comparing the simplicity of his new lung engine with the defects of other engines employed in the steam vehicles of contemporary pioneers, it occurred to him that his ultra-light and compact engine

> was well adapted to sustain the concussions to which such a machine must necessarily be exposed.

Like Thomas, he proceeded slowly and methodically. First, he built a scale model which promised well but failed to produce enough power when scaled up to full size. Undismayed he wrote:

> When once the mind has been exercised towards a certain point, it is no easy matter to apply it in a different direction.

So, he entered upon a course of investigation into the desirable properties of boilers—no easy task, as to be useful they had to comprise three apparently contradictory properties. They had to be capable of producing great quantities of high-pressure steam while still being safe to use without causing any damage should the boiler burst, and they had to be of small compass. Nobody else had been able to do this or the roads would already have been crowded with steamers. This was the barrier that needed to breached if steam on the roads was to progress. It took him three whole years of constant experimental manufacture before, in 1827, he found what he was looking for. His experiments are described in his *Narrative*, which was published in 1838 (Fig. 3.5).

NARRATIVE

OF

TWELVE YEARS' EXPERIMENTS,

(1824—1836,)

DEMONSTRATIVE

OF

THE PRACTICABILITY AND ADVANTAGE

OF EMPLOYING

STEAM-CARRIAGES

ON

COMMON ROADS:

WITH

Engravings and Descriptions

OF

THE DIFFERENT STEAM-CARRIAGES CONSTRUCTED BY THE AUTHOR.
HIS PATENT BOILER, WEDGE-WHEELS, AND OTHER INVENTIONS.

[STEAM PHAETON.]

BY WALTER HANCOCK, ENGINEER.

London:

PUBLISHED BY JOHN WEALE, ARCHITECTURAL LIBRARY,
HIGH HOLBORN:

AND J. MANN, CORNHILL.

1838.

Fig. 3.5 Title page of Walter's *Narrative*.

Fig. 3.6 Walter's original three-wheeler steam vehicle.

With his boiler built, he then turned his attention to designing and building a vehicle in which to put it. He was working, as all pioneers must, in completely uncharted waters and he soon found that much had been published by impractical theorists, which was to mislead him. It had, for instance, been suggested that the metal rims of wheels would not give sufficient bite on the road surface to give traction, so that to move a vehicle along it would require some form of walking apparatus by means of levers acting on the road. This he would eventually find to be unfounded, but it cost him much time and money.

His first machine was a small experimental test bed (Fig. 3.6) and Walter soon found that it incorporated many deficiencies which were mainly due to lack of power from the engine, but he noted, writing as always in the third party:

> It ran for many hundreds of miles in experimental trips from the writer's manufactory in Stratford.... In every instance it accomplished the task assigned to it, and returned to Stratford on the same day it set out.

From his earliest excursions he chose deliberately to drive in full view of the public in the belief that no success could be achieved unless the public mind had been persuaded of the safety and reliability of steam propulsion on the common roads. In spite of all its faults, this little prototype convinced him that a steam-powered vehicle was now a practical proposition and popular acceptance could soon be hoped for.

The next question that arose was what sort of vehicle was the most likely to further his cause? It was an expensive business constructing a

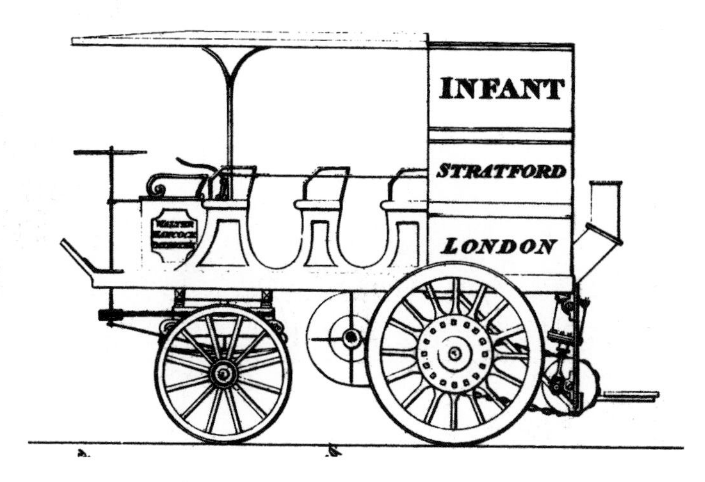

Fig. 3.7 The Infant.

self-propelled coach and whereas there were ample facilities for the horse-drawn coach already in place there was none for steamers. Stations would have to be set up for the provision of coke and water and this was also likely to be expensive. They would have to be manned and maintained. Then, as nothing was likely to succeed unless the public accepted the principle, what better than a vehicle that carried passengers, as their very presence would indicate that they were confident they would survive the experience. It was a brave idea for any failure would be horribly public.

His first vehicle, a compact omnibus seating 10 passengers, he modestly named the 'Infant' (Fig. 3.7) as that was the state of the art at that time and he perceived it would soon grow into a healthy adult. If the mechanical challenges were great, he now found the prejudices of the public even greater.

In these early experimental trips the writer experienced the usual fate of all those who run counter to long standing usages and prejudices; namely to be ridiculed by the many and encouraged by but a very few, and fiercely opposed by all whose personal interests were threatened with injury by his proceedings. The popular mind had not yet become sufficiently familiarized to the notion of disposing with horses in common road traveling. If requiring temporary accommodation through the failure of some part of the machinery—a circumstance naturally enough of frequent occurrence in this early period of his locomotive career, he usually experienced the reverse of kind or considerate treatment. Exorbitant charges were made for the most trifling services, and

Fig. 3.8 An imagined view of White Chapel Road in 1830 (published in 1828 in 'Alken's Illustrations of Modern Prophecy').

important facilities withheld, which it would have cost nothing to afford. If, again, he happened to be temporarily detained on the road from want of water, or from any other cause, he was assailed with hootings, yellings, hissings, and sometimes even with the grossest abuse. It is true, this latter description of treatment proceeded chiefly from the rabble; but he regrets being obliged to add, not exclusively so. Great obstruction was also continually experienced on those occasions from waggons, carts, coaches, vans, trucks, horsemen, and pedestrians, pressing so close on the carriage, as sometimes to preclude the possibility of moving. Altogether the writer's situation was in general any thing but agreeable; often most irksome and irritating, sometimes very hazardous.

The design of the Infant proved excellent and the vehicle met all Walter's expectations. He was able to write:

The feelings of the writer . . . fully compensated him for all his previous annoyances. He had dissipated the doubts of friends, and disappointed the anticipations of enemies; he had conquered difficulties before deemed insurmountable, and placed the power of steam, in comparison with that of horses, in the most

Fig. 3.9 An imagined view of Regent's Park in 1831 (published in 1828 in 'Alken's Illustrations of Modern Prophecy').

advantageous position. Assured that he was approaching towards complete success, he remodeled the entire arrangement of the machinery . . . and such other alterations and improvements adopted, as had suggested themselves during actual work upon the road. No ways disheartened by any of these untoward circumstances, the writer persevered in his experiments, and as the novelty of such exhibitions wore off, so also did the opposition which they at first produced. Clearer sighted views and kindlier feelings began gradually to prevail; more serious convictions of the practicability and advantages of substituting inanimate for animate power in common road traveling, and greater readiness to promote by word and deed the success of the project.

Nevertheless, these views and feelings were not always those of some cartoonists whose portrayals of life to come in a world of steam locomotion would make even the stoutest hearts quail (see Figs 3.8, 3.9 and 3.13)!

The Infant had one more task to perform. It was widely held that no steam vehicle could possible ascend a hill without propellers so, to

Fig. 3.10 The Era.

overcome this final prejudice, Walter organized a public demonstration on Pentonville Hill, which had a gradient of 1 in 18.

A severe frost succeeding a shower of sleet had completely glazed the road so that horses could scarcely keep their footing. The trial was therefore made under the most unfavourable circumstances possible: so much so that, confident as the writer felt in the powers of his engine, his heart inclined to fail him. The carriage however did its duty nobly. Without the aid of propellers or any such appendages... the hill was ascended at considerable speed, and its summit successfully attained, while his competitors with their horses were but a little way from the bottom of the hill.

In February 1831 he commenced running the Infant regularly for hire on the road from Stratford, Essex, to London:

not certainly with any anticipation of profit, although the writer's outlay at this time... had been considerable, but as a means of dissipating any remaining prejudices and establishing a favourable judgement in the public mind as to the practicability of steam traveling on common roads. It is an undeniable fact, and a source of proud satisfaction to the writer that a steam carriage of his construction was the first that ever plied for hire on the common road, and that he had achieved this single handed.

Not surprisingly speculators, for this was the age of speculation, began to take an interest in these ocular proofs of steam on common roads. Walter wrote:

The writer has now arrived at a period in the history of his locomotive career, when it became rather chequered by an association with other individuals.

Fig. 3.11 Enterprise.

His first approach came from a group of company promoters who styled themselves the London and Brighton Steam Carriage Company. They ordered a carriage that Walter named 'Era' (Fig. 3.10) as it seemed that the new era of steam traction had just dawned. Demonstration journeys were undertaken in 1832 to Windsor and elsewhere; the carriage performing flawlessly but the proprietors had been unable to attract sufficient investment; interest waned, and the new carriage stood idle in the yard at Stratford.

Further approaches now came from another group styling themselves the London and Paddington Steam Carriage Company. This time Walter was a bit more cautious and, while accepting the order for the new carriage, demanded and got a deposit and an agreement for stage payments. He was learning. He was assured that if the trials were satisfactory he would receive orders for a further two carriages.

Once more the dawn of steam seemed to be breaking. The new carriage, 'Enterprise', was completed by January 1833, when it was delivered to the company for final painting in its new livery (Fig. 3.11). The magnificent machine in maroon and gold made its appearance on the road in April 1833 with Walter in the driving seat, and it ran a regular service between the City and Paddington for 16 days without mishap. A shareholder wrote a promotional puff in the *Mechanics Magazine* about how well the trials had gone, what the fares would be and how, after the delivery of the next two carriages which were on order, they would instigate

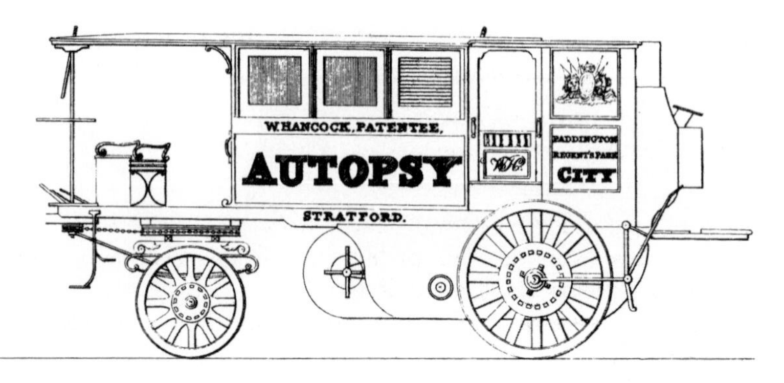

Fig. 3.12 Autopsy.

a regular bus service with each carriage performing 14 journeys a day. It all sounded very good but it was not. It implied that Enterprise had been built in the company's own yard in Charles Street and the next two carriages would be built there also. It did not take long for Walter to discover what was really going on. The final payment was not made for Enterprise, despite repeated demands. Walter demanded to know why and vague excuses were proffered. He then asked for the return of the carriage and this was refused. While this smokescreen of prevarication was being laid, Mr Redmund was busy stripping down Enterprise so he could clone it for his company. Eventually, Redmund's machine came out of its yard for a public demonstration. They had been able to copy every part of it except, crucially, the patented boiler, which they dared not imitate. Now the quality of Hancock's carriage became clear. As Redmund's machine trundled along the road, first steam began to escape in clouds, then water began to pour out of every joint until the engines were deprived of both and thereupon it came to a stop, much to the enjoyment of the crowds who had come to witness it. They expressed their hearty derision in the usual way and the sounds of the hootings and hissings could be heard long after the broken behemoth had been towed back ignominiously to its yard. It had all along been a planned deception and it had failed. However, the cost to Walter had been high. He wrote:

> ...the whole project however met with the fate it deserved, in a total and complete failure, alike disgraceful to the morals, as to the pretensions of the parties involved.

Although this sequence of events cost Walter severely in financial terms his reputation was, if anything, enhanced and he was soon at work building the unusually named 'Autopsy' (Fig. 3.12). The writer Baudry de Saunier[5] gave his views on the name when he wrote:

it is a queer way to encourage people to try a new method of transport, which invokes the image of the operation which they might be unfortunate enough to suffer.

But the name is derived from the Greek and can most simply be translated as 'see for yourself'—an apt description from a medical viewpoint and from Walter—'see for yourself what I can deliver!'

Autopsy made a successful visit to Brighton before plying for trade between Finsbury Square and Pentonville, but after a month Walter withdrew the service because of

the want of suitable premises and stations for the store of coke, supply of water, standing for the carriages &c., superadded to the necessity of the writer's daily personal attention at his manufactory at Stratford.

The problems of fuel and water supplies which had been noted earlier were again imposing limits on the practicality of a steam carriage service while Walter's own reluctance to train a driver, thus leaving him more time at his manufactory, should have made the eventual failure of the project inevitable.

Fig. 3.13 '[A] few small inconveniences—(Locomotion)—There's nothing perfect' Published by Thomas McLean, 28 Haymarket (undated).

However, fate was soon to take a hand and not only remove the future of steam carriage transport on the common road from Walter's carriages but also from those of all his competitors. The first event in April 1834 was a disastrous fire at Thomas' works in Goswell Mews, which hit Walter from two directions; first, he would lose his regular income from supplying Thomas with all the engineered components he had been manufacturing for him, and, second, he would have to turn his attention to rebuilding the machinery lost in the fire and surely, he thought, redesigning and strengthening it to meet Thomas' ever-growing workload. Any work or driving related to his own project would have to be put on hold.

The second event was a most serious steam carriage accident which occurred in Glasgow in August of the same year.[6]

It appears that the carriage, having gained the summit of the acclivity at the place in question, stopped for a minute or two to take in a supply of fuel and water, when, just as it was in the act of proceeding on its journey, one of the right-hand wheels gave way—the machine came to the ground with terrific violence—the boiler was instantly crushed as flat as a pancake, and simultaneously with which, the bottom of the vehicle was shattered to atoms by the explosion, and all the passengers, twelve in number, were more or less injured. Among the sufferers, Captain E. B. Gilmer, father-in-law to P. A. Black, Esq., of this city, sustained a severe concussion of the brain, which put a period to his existence, on Tuesday night, at 6 o'clock. Mr. Thomas Blackwood, traveller for Messrs. White, Urquhard & Co., also expired on Wednesday morning, at 8 o'clock, in consequence of the injuries he had received. Mr. James Morrison, merchant, Gallowgate, had his thigh severely fractured, which caused his death on Wednesday, about 3 o'clock afternoon. Mr. William Sym, miller, Partick, who got his arm broke besides several internal injuries, also died yesterday afternoon at five o'clock. Mr. James Sargeant, merchant from Leicester, had also his thigh broken; the limb was immediately amputated...

In point of fact there had been no explosion and those who had perished had died from injuries received when they threw themselves from the vehicle in anticipation of one.[7] The remaining victims suffered from the effects of scalding steam and flying red hot cinders, but it was enough. The speculators, who were now trying to interest the public in railway shares, cried 'Explosion', which the press amplified to make a better story and the public believed it. This left the future of steam propulsion on common roads badly, perhaps even fatally, injured and turned the populous towards the steam locomotive, with its defined

Fig. 3.14 Walter Hancock, a painting by Charles Hancock 1835.

path and separation from other forms of transport, including the innocent pedestrian bystander.

Walter withdrew both his carriages from public service in November 1834 as Thomas resumed his business in the rebuilt factory and worked overtime to catch up with his delayed business. Walter had to follow suit making the various valves and other metal fitments. Over the New Year, he was invited to take the Era to Ireland to demonstrate its capabilities. Renamed 'Erin', in honour of the Emerald Isle, it performed faultlessly and returned to Stratford after two weeks. No orders were forthcoming and Walter now accepted that his dream of steam carriages running on the common roads throughout England would never be realized. Although he was later to run a service from the City to Paddington, putting all his carriages on the road for five months, and although he was to build an even larger and more powerful carriage called 'Automaton' (see Fig. 5.7, page 71), he did this more in a spirit of defiance in the face of the duplicity and stupidity of the world than with any hope of continuing to promote his vision. The world was not yet ready for he was half a century ahead of it.

The year 1834 had been testing for Walter (Fig. 3.14). He was exhausted and no doubt depressed by his failure to spark any commercial interest in his vehicles. His household was run by Rebecca Wendon, who was 31 and very fond of him. As he sat gloomily casting round for new avenues one thing led to another and in October 1835 a daughter was born to Rebecca. She was christened Elizabeth Ann Hancock Wendon, 'by desire of the parents.'

4

Life, Death, and Bankruptcy

Charles becomes a successful painter—reputation grows—makes lithographic sketch of William Corder (the Red Barn murderer). William becoming successful—cabinet for King George IV but still not making money—wife opens shop but to no avail—the family returns to London and William gifted one of Thomas' patents to set him up as a bookbinder—business fails and second bankruptcy follows. John now in his own rubber business and thriving while Thomas hears of Charles Macintosh and his waterproof materials—Macintosh's background—involvement with the Birley brothers of Manchester. Thomas takes licence from Macintosh—discovers rubber solutions made with his masticated or pickled rubber superior to Macintosh's and give better products—offers to cooperate with Macintosh—eventually an accord reached. Thomas builds a 'turn-key' factory for two French entrepreneurs but still keeps pickle secret. John has consumption, sells out to Macintosh and moves to Cornwall. Thomas' factory burns down. John dies—family returns to London—his widow suffers from acute depression and is unable to cope. Thomas becomes partner in Macintosh's company and, a confirmed bachelor, takes responsibility for John's nine children.

We must now leave Walter's story for the moment and catch up with Charles.

Following the unfortunate debacle of the anti-fouling process, Charles returned straightaway to his painting. Taking rooms at Mrs Connings in the Market Place at Reading in June 1823, he advertised[1] himself as 'portrait and miniature painter'. As the months passed he added 'landscape and animal painter' to his description, and by 1826 he had become a 'Member of the Norwich Society of Artists'. He was clearly working very hard at his easel, taking pupils one day a week and trying to sell his paintings from his lodgings. By 1826 he was also advertising[2] that he offered for sale 'Hancock's scentless water colours':

The novel and elegant method of preparing these colours damp, in Hancock's patent India rubber bottles gives them every advantage and facility in use. The bottles being entirely impervious to air, will not only keep the colours in a moist state for any length of time, but enables the artist to supply his palette without the trouble of continually mixing up his colours.

He also offered 'Hancock's scentless oil in which he mixes colours for sale, for painting after the manner of oils'.

Catalogues of artists and their works for the Royal Academy Summer Exhibition as well as for the Royal Society of British Artists and the British Institution show that hardly a year went by when he did not have a painting hung by one or more of them. By now he had become a truly accomplished professional artist, but earning a living was still a struggle. In the ledgers of Ward and Merriman, bankers, in Marlborough, there is an entry[3] showing that he borrowed £40 in 1825 and that Mrs Hancock, presumably his mother, made two repayments of £10 each in the same year. It is not surprising then that if a journalistic opportunity arose he would grab it. When the infant prodigy Master William Grossmith played the part of the infant Roscius in Reading in 1824, he painted the famous scene and had an engraving made to take advantage of the local interest.

Later, in 1828, he was to travel to Bury St Edmunds to attend the trial and execution of the infamous William Corder, the Red Barn murderer. Many artists made sketches of Corder but Charles Hancock took the trouble to have his sketch authenticated by the trial and prison authorities as 'the most lifelike they had seen',[4] thus giving his lithographic print an edge over all the others. After the hanging, Corder's body was dissected for medical research and his skeleton cleaned and given to the West Suffolk General hospital for teaching purposes. Charles took the opportunity of making two further lithographic sketches of the hospital to sell to those eager to have a memento of this grisly event.[5] While staying with William in the Butter Market, Charles managed to find time to paint a portrait of William since he, despite his earlier bankruptcy, had been making something of a name for himself as a cabinet-maker (Fig. 4.1).

On 19 October 1825 the *Bury and Norwich Post* reported that:

Mr Hancock of this town has just completed an elegant cabinet (made of elm) for the King's palace at Carlton House. In workmanship and beuty [sic] of wood, which was from near Sudbury, it has been very justly admired by a number of distinguished visitors at our fair last week.

One of the distinguished visitors must have been the Marquis of Bristol, who then lived nearby at Ickworth House, since among his

Fig. 4.1 William Hancock painted by Charles Hancock.

possessions resides a confidential letter box with a secret locking mechanism made by William. Nine months later, on 19 July 1826, the newspaper reported further on William's gift to George IV.

We noticed some time ago the preparation of a beautiful piece of furniture of British elm by Mr Hancock of this town, for the purpose of being offered to his majesty for his acceptance, and we are glad to find that Mr H has received the gratifying return of a draft for 100 guineas, enclosed in the following[6]

Carlton House, 10th July, 1826

Sir, I am commanded to acquaint you that the King has been graciously pleased to comply with your desire, and to accept the cabinet prepared by you, and forwarded to Carlton House, for the purpose of being presented to His Majesty. His Majesty is gratified with your dutiful attention and has further commanded me to transmit to you a draft (which I enclose) for 100 guineas, as a mark of his Majesty's gracious approbation:

I have the honour to be Sir, your very obedient servant,
Thos. Marrable

The description of the cabinet in the royal collection reads:[7]

A cabinet of superior beauty and workmanship made of British elm it is of upright square form 5ft 7½ inches high, 3ft 10½ inches wide, by 1ft 3¾ inches deep. The cabinet has folding doors, with a column to each door—the outside of the cabinet is formed into long pannels [sic] with carved and gilt leaf mouldings—the upper part of the inside of the cabinet forms a recess with looking

Fig. 4.2 William Hancock's elm cabinet. The Royal Collection © 2008, Her Majesty Queen Elizabeth II.

glass back and sides with four columns, and four half columns. The columns have gothic arches to them. At top the columns are multiplied by reflection in the glass. A secret spring slider under the columns—the centre of the internal part is fitted with eight sliding shelves lined with red velvet—and three drawers the wings of the cabinet is [sic] of circular form—with seven drawers for papers on either side—in the centre of each wing is a recess with doors—the recesses contain six small drawers each—the doors of the recesses are covered with crimson velvet and gilt ivy wreaths and spears.

This cabinet was presented by Mr Hancock to His Majesty as a specimen of what British wood may be brought to—and its perfections showed—if manufactured under the hand of a skilled workman.

William, ever the traditionalist, no doubt intended that his cabinet (Fig. 4.2) should turn the King's taste away from the foreign influences to which he was so famously addicted and back to the best of British craftsmanship. He may have been assisted in this design by the Marquis of Ailesbury, who was at that time was Master of the Royal Household and not unacquainted with the Hancock family as he had been James Hancock's landlord in Marlborough. (The Marquis was soon to commission Charles to immortalize two of his favourite horses and hounds.[8])

William leaves us in no doubt that he is a superb craftsman, but still he struggled to make a living and to feed his rapidly growing family. By

Fig. 4.3 William's business card.

5 January 1831 the finances of the family must have reached crisis point again as, on 5 January, the local newspaper carried the following:

New tobacconists shop just opened at the corner opposite The Greyhound, Butter Market, Bury. MRS HANCOCK begs to acquaint her friends and the public generally that the increasing expenses of a numerous small family have induced her to open the above shop (the adjoining one to her husband's cabinet and upholstery warehouse). Fancy and other snuffs, Tobacco, Fine flavoured Segars, genuine teas coffees etc. in which business she humbly solicits patronage, assuring them of supplying them with good articles at moderate prices, and of her grateful attention to their commands. NB publicans and wholesalers supplied at the usual allowance. For the accommodation of her customers Mrs Hancock will have The Times newspaper on the counter daily.

Not even the King's endorsement and the extra income from the shop enabled them to earn a living so, bowing to the inevitable, William chose to return to London to join Thomas in Goswell Mews, where he was handed the latter's patent method for book-binding, a process now known as perfect binding. He set up business in Glass House Yard, Aldersgate Street (see Fig. 4.3). Craftsman he undoubtedly was but businessman he was not! Perhaps his heart just was not in this sort of commercial enterprise, or his domestic and family responsibilities were overwhelming, for by 1839 he was bankrupt once more

While their younger brothers struggled in their own fields between 1820 and 1830, Thomas and John (Fig. 4.4) ploughed their own profitable and independent furrows. John had made a speciality business of turning rubber sheet into rubber pipes and hoses. Soon he was to be found laminating various fabrics in combination with rubber solution

Fig. 4.4 John Hancock.

to form industrial and reinforced hoses. These found a ready market with the brewers,[9] who were then using stitched leathern hoses which leaked copiously and wasted much beer around the brewery. John had discovered that by inserting a coiled wire into his laminated hoses, they would resist pressure by suction, and this also appealed particularly to brewers. The Norwich Union Fire Insurance Company quickly adopted Hancock hoses for their company fire engines.[10] Walter also made use of these hoses, which he coupled to his various steam carriages to suck water from ponds and reservoirs. It was a growing and thriving business which necessitated, in 1832, opening a retail warehouse in London (see Fig. 4.5).

Meanwhile, Thomas had come across the work of Charles Macintosh and his patent of 1823 (No. 4804) for 'Rendering Fabrics Waterproof'. A process which at first glance seemed alarmingly similar to the one he was using.

Charles Macintosh[11] was 20 years older than Thomas, having been born in Glasgow on 29 December 1766 (Fig. 4.6). His father owned a dyeing business but from early in his life Charles went his own way, opening the first alum works in Scotland in 1797 and, with Charles Tennant, developing a dry bleach made from chlorine and slaked lime. This made him a considerable fortune and enabled him

CAOUTCHOUC.—JOHN HANCOCK, the Original Manufacturer of Articles of DOMESTIC and Medical Use in CAOUTCHOUC, or INDIAN RUBBER, respectfully informs the Public, that he has opened a Warehouse, (the only Establishment of the kind) at No. 11, Agar-street, Strand, opposite CHARING-CROSS HOSPITAL, where all his Manufactures in Caoutchouc may be had WHOLESALE and RETAIL.

☞ A Descriptive Prospectus GRATIS.

Fig. 4.5 Advertisement in *John Bull* 30 September 1832.

Fig. 4.6 Charles Macintosh.

to research into various other fields of chemistry. By 1792 Glasgow had begun to introduce gas lighting into the streets as well as into certain prestigious properties, and in 1817 the Glasgow Gas Light Company was formed. The gas was produced from coal, and in 1819 Macintosh contracted to buy the waste products as he could extract ammonia from them which, in turn, could be used by his father's company to make a violet-red dye called cudbear. He was then left with yet another waste material—a complex mixture of chemicals, not dissimilar to today's petrol or gasoline, which was known collectively as coal tar naphtha or coal oil.

This substance had been noted by Fabrioni in Italy as early as 1791 as an excellent solvent for rubber, but it was not until 1818 that James Syme in Edinburgh proposed that it could have some commercial significance, adding that it was cheap and readily available with the advent of the new gas lighting. Macintosh began experimenting with coal oil as a solvent for rubber in 1819 and found that he could paint the solution on to fabrics, which, when dry, became waterproof—although it was also sticky and gave off a foul smell. His brilliant idea and the heart of his patent was simply to press two sheets of fabric together with the dissolved rubber sandwiched between them.

In 1824, Macintosh persuaded the Birley brothers, cotton spinners and weavers of Manchester, to build a factory next to their mill in which he could manufacture his rubberized cotton. However, he encountered numerous problems, not least the persistent and nauseating smell of his finished products. This difficulty prevented his articles being widely adopted by society, although there was a large and regular demand from the armed forces and merchant navy. No doubt the waterproofing properties of the fabric more than compensated for one extra smell among so many to which they were exposed. Further difficulties were encountered arising from the two basic properties of raw rubber—its twin tendencies to degrade to a syrup-like liquid in the presence of sunlight and air and to become brittle if left to drain dry in a cold outhouse or porch overnight in winter.

In spite of his lack of success with John Clark's patent, Thomas still believed there was a great future in materials coated with solutions of rubber and in 1825 he obtained from Macintosh a licence to manufacture Macintosh's patented 'waterproof double textures'. It took him but a little time to realize that his solutions, prepared from masticated rubber with a mixed solvent of purified oil of turpentine and naphtha (coal oil), were able to have a higher rubber content than those of the patentee. He also observed that they more readily gave a uniform film on the cloth with less penetration through it and, better still, produced far less odour.

Thomas Hancock now took a brave initiative. He wrote to Charles Macintosh and Co. offering the company the benefit of his own rubber solution. It is not surprising perhaps that this offer was at first rejected, for here was Charles Macintosh and Co., a subsidiary of Birley brothers, among the largest cotton spinners in Manchester, being offered a substance by an unknown and unheard of ex-coach-builder from Stoke Newington. The Birley brothers' mill alone had cost several hundred thousand pounds to build and equip, and employed at least 16,000 men, women and children whose wage bill amounted to £40,000 annually. Their annual consumption of raw cotton amounted to over 4 million pounds weight per year. Their steam boilers burned over 8,000 tons of coal annually; their machinery required 5,000 gallons of spermaceti lubricating oil. In one gallery alone there stood 600 power looms. It was a vast enterprise.[12] Within the same site lay the buildings the Birleys had erected for Charles Macintosh and Co. Here, depending on the season of the year, were employed between 200 and 600 hands working in a factory that would within a few years be consuming 250,000 pounds weight of rubber, for which 100,000 gallons of spirits would be needed to prepare the solutions.[13]

Thomas recorded:

I wrote to Mr. Macintosh, and offered to supply Messrs. Macintosh and Co. with my solution; but this offer was at that time declined. In the meantime these fabrics were quietly becoming known to the public, and the goods were taken up nearly as fast as we could respectively produce them. The late lamented Captain John Franklin (after-wards Sir John Franklin), in a letter to Mr. Macintosh dated 30 April, 1824,[14] after acknowledging the receipt of a large quantity of waterproof canvas for covering boats, &c., says, 'Will you also make up four life preservers of a size for stout men, and eighteen bags about six feet long, and three broad, fitted with corks for filling with air for the party to sleep on, and four for pillows of the size of the one you gave me.' I insert this extract to show how early these waterproof double textures were appreciated, and the application of the material to air-beds, pillows, and life-preservers.

The airbeds which Franklin ordered would appear to be simple one-compartment units but Thomas already knew their deficiencies! He commented that:

a bed made in only one compartment, so that, when inflated, it assumed a pin-cushion shape, and its rotundity was such that to place yourself upon it and remain there was impossible; try as you might to balance yourself, in a moment you lost your equilibrium and came rolling on to the floor; each bystander thought he could do it, but the airbed set him tumbling about, and all at length acknowledged a defeat and declared that air-beds 'would never answer.' A portion of the air was let out, but the same kind of objection remained; and although this was repeated until nearly the whole of the air was exhausted, this principle of construction was still evidently defective.

Franklin's airbeds would have been fitted into wooden 'boxes' on board ship which would help to retain the sleeper, but Thomas had already overcome these difficulties:

> It consisted in preparing a case of ordinary bed ticking divided into seven or eight compartments; an air-proof cylindrical chamber of a proper length and diameter was made for each of these compartments, and inflated to any desired degree. This was an obvious improvement, and the air-bed thus constructed, whilst it yielded sufficiently to the form of the body, supplied at the same time a more elastic resting-place than ever the human form had before reclined upon.

Macintosh and Co. had reached its eminence by pursuing a policy of continuing experiment and to them it was soon apparent that Hancock's rubber solution and coated cloth was far superior to their own. The result was a foregone conclusion:

> Late in the year 1825 it was proposed that Messrs. Macintosh and Co. and myself should come to some arrangement by which articles made under Mr. Macintosh's patent should emanate from the establishment of the firm; and, during this correspondence, I stated in some detail to Messrs, Macintosh and Co. the large extent to which I contemplated carrying out my views in regard to the applications of rubber, and the patents I had taken out to secure them, and that it would be necessary for me to engage with capitalists ready and willing to cooperate with me. This correspondence resulted in an arrangement in February, 1826, by which I engaged to manufacture for Messrs. Macintosh and Co. the articles covered by Mr. Macintosh's patent, providing that a partnership should be avoided (to which I ever had a dislike), that my name should be stamped on all the goods I made for them, and that no other goods but those made by me should be sold within the limits of the bills of mortality. This arrangement did not interfere with my business in other respects; for the present our relations extended no further.

It is interesting to note with what caution Macintosh and Thomas Hancock gradually approached each other and how their future relationship was born out of a growing mutual respect and friendship and not out of an urgent desire for unlimited wealth. The rubber solution made at Goswell Mews was not yet being supplied to Macintosh's works in Manchester, and he must have been intrigued as to how the Hancock-made cloth could be so clearly superior to his own.

It had never been the intention of either man to enter the garment manufacturing and retailing trade but they were forced into it by the attitude of the tailors who made up the garments for their clients (see Fig. 4.7). Each seam was, of course, stitched and this offered a perfect route by which water could pass through and soak the wearer. Although the tailors were advised to send the finished articles back to the factory

Fig. 4.7 The shop of Charles Macintosh & Co., Charing Cross, 1840.

for proofing, most were offended and refused to cooperate. As Thomas was to remark:

> Some of them persisted, and actually made a double row of stitches to make sure work of it.

Even in their own tailoring establishments it proved difficult to teach the tailors new tricks and many was the time when a leak was put down to a pin or needle hole in the body of the cloth.

Once again, Thomas' ability to meticulously observe the problems of different modes of water penetration and to identify the causes, in some cases

> tracing back the defective articles to the pieces from which they had been cut, and then to the mill books, to ascertain if possible if any deviation from the usual course could be discovered

made all the difference between the survival and collapse of the whole business. Traits shared by both Hancock and Macintosh were honesty and fairness to all their clients and full refunds were given to anyone

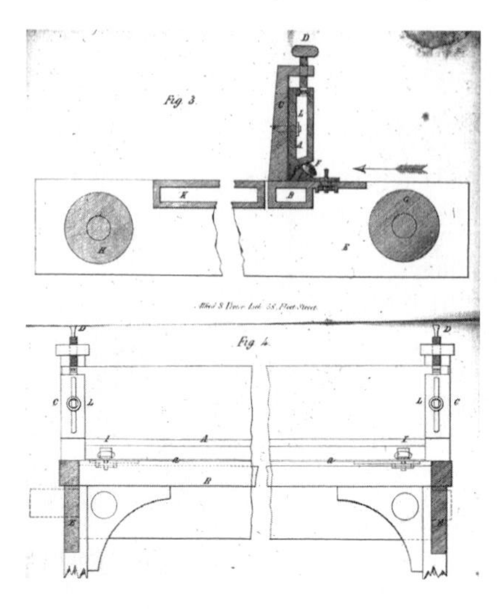

Fig. 4.8 Walter's design for a spreading table.

who was not content with their purchase. Not surprisingly, the business was yet to show a profit.

The clause in Thomas' agreement with Charles Macintosh which ensured that he could carry on his own business while working with the latter was soon to be of particular importance. In 1828 he was approached by Messers Rattier and Guibal of Paris, who wished to establish the first rubber product manufacturing company in Europe and who asked whether he would be interested in supplying the equipment for it. This posed a problem for Thomas as he had still not patented his masticator and was most anxious to keep it secret. He discussed the matter with Walter, who would have to design the machinery, and they concluded that the project could go ahead if the Frenchmen were content to be supplied with Thomas' masticated rubber rather than being given the knowledge and machinery to prepare it for themselves. This was agreed and Thomas trained some additional workers, put them under the care of Edward Woodcock, a trusted friend, and sailed to France to set up what may have been the first example of a 'turn-key' factory in Europe. Edward Woodcock was to remain with the company until his retirement.[15]

By 1829 Thomas had become firm friends with Charles Macintosh and although he still refused to consider a partnership he was prepared to extend his relationship with Macintosh by agreeing to supply

Fig. 4.9 An early spreading table (circa 1852).

masticated rubber, dry, from London so that solutions could be made at the Manchester works. There the rubber solution had been spread over the fabric with paint brushes, but Thomas, no doubt on Walter's advice, introduced a spreading table with a trough to hold the solution, together with an adjustable blade by which the thickness of the solution could be controlled, according to its viscosity (Figs 4.8 and 4.9).[16] This design remained unchanged for many decades, a lasting tribute to its designer. The arrangements between Hancock and Macintosh were now running very smoothly and as trust grew between the two men, and so did the profits of the company: a happy state of affairs.

Both men must also have been gratified that the religious and humanitarian principles on which they had built their lives and businesses were, to a small extent, being followed by the government of the day. The years of 1832 and 1833 saw the introduction of the first Reform Bill, the abolition of slavery, the first Factory Act (which limited child labour although not in rubber factories) and the first stirrings of state-funded education. However, although the commercial sky was blue, a cloud no bigger than a man's hand was forming on the horizon. John had begun to cough. As this became worse, it was soon only too obvious that he was suffering from consumption and the disease was quite advanced. He now had a large family in Fulham, nine children

Fig. 4.10 Humphrey Davy's house Varfell, in Ludgvan, Cornwall, where John stayed.

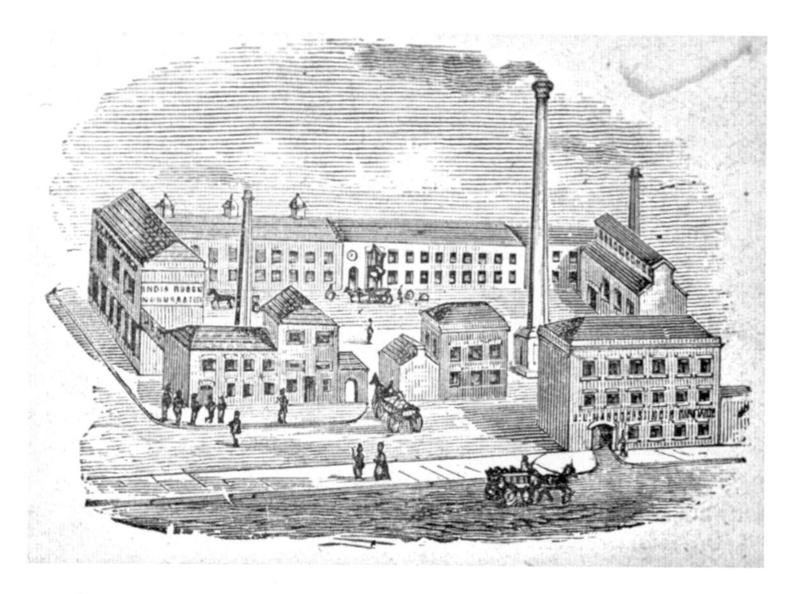

Fig. 4.11 The new layout of Thomas' factory in Goswell Mews (as it was in 1850).

in all aged between 18 and five, and a wife, Fanny Maria, who was already suffering from depression, a condition to which she was prone. What was he to do? As his own condition deteriorated, he decided that his only option was to take his doctor's advice and remove himself to Cornwall, where the sea air would be better for him—it was his only hope of improvement and might give him a few more years. Here he was fortunate, for almost certainly through the good offices of Michael Faraday he was generously offered the late Sir Humphry Davy's house and laboratory in Ludgvan, which was near Marazion (Fig. 4.10). In 1833 he moved there with his whole family, and devoted himself to developing safety fuses for the tin miners[17]—to the last dedicating his remaining time to a worthy humanitarian cause.

His business in London was sold to Charles Macintosh & Co. and moved from Fulham to Goswell Mews, but within a year it had burnt to the ground. Given that the men were working by candle light and were surrounded by highly flammable solvents it was perhaps surprising that this had not happened before but fortunately there was no loss of life and, as Thomas was to write:

> The premises and stock being insured, my loss was not very great. In laying out my plans for the new works, I took care to have the buildings detached, with a considerable space between each of the three, and without windows or doors on opposite sides. These precautions and some others enabled me to get them insured again.

His detailed calculations for the insurance company showed his losses to total £1081-9s-9d, which he obligingly rounded down to 'say £1000'.[18]

The fire gave Thomas the opportunity to redesign his processes and rationalize both them and his relationship with Charles Macintosh & Co. It was decided to rebuild Goswell Mews (Fig. 4.11) solely as a fabricating plant and

> In the new buildings no provision was made for manufacturing either rubber or solution, which was now supplied from Manchester; this alteration afforded me much greater space for spreading machines for water-proofing and air-proofing cloth, and for the manufacture of air-proof articles, proofing the seams of garments, and a variety of other things, the whole business having now greatly increased in magnitude.

With the reorganization completed, the process of integration which had been growing between Thomas and Macintosh together with their mutual respect and affection for each other would finally tempt

SACRED TO THE MEMORY OF
JOHN HANCOCK
BORN AT MARLBOROUGH, WILTS, 28 FEB. 1788,
DIED 13 MARCH 1835, AGED 47,
AND WAS BURIED IN THIS CHURCHYARD.
HE SPENT MANY YEARS, FROM 1826, AT FULHAM, LONDON,
IN APPLYING THE DISCOVERIES OF HIS BROTHER THOMAS,
IN INDIA RUBBER, TO SURGICAL AND OTHER PURPOSES
AND RESIDED IN THIS PARISH THE LAST 2 YEARS OF HIS LIFE
TO APPLY IT TO SAFETY FUSES FOR MINERS.
HE COMMITTED HIS YOUNG FAMILY OF NINE CHILDREN
TO THE CARE OF HIS UNMARRIED BROTHER THOMAS
WHO RECEIVED AND PROVIDED FOR THEM AS HIS OWN.
THOMAS HANCOCK
DIED IN 1865 AND WAS BURIED AT KENSAL GREEN, LONDON.
THIS TABLET IS ERECTED BY THE SURVIVING CHILDREN
IN MEMORY OF THEIR DEAR FATHER AND UNCLE
AND IN DEEP THANKFULNESS TO ALMIGHTY GOD.
WHO REMEMBERED US IN OUR LOW ESTATE. PS. CXXXVI. 23.

Fig. 4.12 Plaque to the memory of John Hancock in Ludgvan Church, which also commemorates his brother Thomas.

Fig. 4.13 The new trademark of Charles Macintosh & Co. showing the influence of Thomas Han(d)cock.

Thomas to overcome his misgivings and give serious thought to accepting a partnership. It would not be long before events were to take a turn that would finally convince him that a partnership was in everybody's best interests.

John died in the spring of 1835. He is buried in Ludgvan churchyard, where in later years his family erected a plaque in the church to his memory (Fig. 4.12). The family returned to London by paddle steamer, battened down below the hatches as the weather was wild,[19] to face an uncertain future. Fanny Maria Hancock never recovered from this final blow of fate and as her depression grew ever deeper it became clear that she could no longer look after her large family. A place was found for her in the Countess of Somerset's almshouses in Froxfield near Marlborough, where she lived, separated from her children and suffering repeated bouts of illness, until finally she had to be placed in a lunatic asylum near Devizes.[20]

The blessings of the new financial security bestowed upon Thomas by the partnership (see Fig. 4.13) came at an opportune moment, for John had commended his young family to Thomas' care; all nine of them. Thomas had seen first hand the poverty and starvation brought about by the war with France followed by disastrous harvests in the years following the eruption of Mount Tambora and was determined to protect his new family from the hardship which was still rife throughout much of the country, again suffering from a series of poor harvests and the import restrictions imposed by the Corn Law.

Whatever was he to do with them?

5

The Family, the Law, and the End of a Dream

Thomas' nephew, James Lyne Hancock, runs Thomas' factory in Goswell Mews while Thomas devotes more time to the Macintosh factory in Manchester—acquires Marlborough Cottage in Stoke Newington for his new family—children educated by Joseph Burrell. Thomas and Macintosh involved in first of several patent battles—this with Everington & Ellis over Macintosh's 1823 waterproof fabric patent—legal arguments presented—jury finds for Thomas and Macintosh. Thomas decides to patent his masticator and spreading machinery. Manchester factory gutted by fire but quickly restored to full functionality. Walter builds his largest steam carriage yet almost for pleasure as public interest collapses—his observations as to why steam road transport was doomed. His last journeys—the sale of his factory.

Thomas Hancock was now, in 1835, nearly 50 and a confirmed bachelor. His rebuilt works at Goswell Mews was running smoothly, now increasingly under the guiding hand of his nephew James Lyne Hancock, while Thomas' principal business interests lay in Manchester. He does not seem to have had many close friends save those in business, and the majority of those were now in either Scotland or Manchester. It is to the latter that we must now turn. Even to Thomas, who was used to travelling about London, the sheer scale of business in Manchester must have impressed him. Unlike London, it was a city that was built for manufacturing on a colossal scale. No one could fail to be impressed by the towering mill buildings or the size and nobility of the public buildings, which had all been provided from the fruits of business and were devoted to the improvement of the education and health of the population of over 300,000. No less impressive were the habits of the proprietors of the great mills. Love and Barton[1] remarked:

The habits of Manchester men of business are marked by the utmost perseverance and energy. Unlike the merchants of London...they permit little relaxation to be associated with the stern demands of business...nothing but sickness is allowed to interfere with a daily attendance to the business of commerce or manufactures. And this attention is not for a few hours only in the day; it commences early in the morning, and is protracted to a late hour in the evening. It is a common thing to see the leading merchants of the town—some of them possessed of wealth to an amount of a quarter of a million sterling—posting from their country villas to their counting houses between eight and nine o'clock in the morning, and many of them do not return home (except to a hasty dinner) till nine or ten in the evening. Business becomes a habit, and this habit becomes a pleasure; and on this account—more than from mere love of gain—they are impelled to proceed onwards in a vocation, which in its enterprise and excitement, presents to them the greatest of earthly charms.

This picture so exactly mirrored Thomas' life, which had, so far, been solely dedicated to his business that he felt truly at home there, but now he had more than just himself to consider and we may only guess at how he viewed his new familial responsibilities. He certainly knew where his Christian duty lay and he was not one to shirk it. After his brother's death, he took sole charge of the children as though they had been his own. Maria and Frances, now aged 19 and 17 respectively, would have to help, playing the role of mother and nursemaid to their younger siblings. Thomas would see to their spiritual, moral, and educational needs.

His first priority was to find a suitable home for his newly acquired family. Just to the north-east of Clerkenwell was the parish of Stoke Newington, which he knew well, having lived there since the 1820s, and which had a reputation for being popular with the Dissenters. It was rapidly expanding, as it provided good houses in an attractive setting within easy reach of commercial London. Merchants, bankers, and stockbrokers spread northwards from Church Street towards Woodberry Down and Seven Sisters Road along Lordship Road and Green Lanes. However, in the 1830s the northern part of Stoke Newington was still mainly rural, dominated by the twin reservoirs built in the 1820s adjacent to the 'New River', which was actually a man-made cut, its erratic route following the contours of the land.[2] Some small-scale development was taking place as indicated in Fig. 5.1 and this gave Thomas the opportunity to acquire a long lease on the property identified as 'plot 60' in Fig. 5.2—the northernmost house of the block of six to the south of Seven Sisters Road.[3] He was also able to take a lease on the meadow to the east of the house (plot 61), the area of which was rather less than 1 acre.

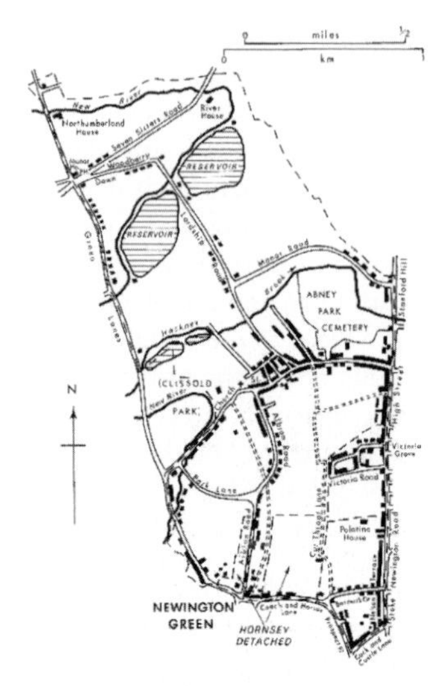

STOKE NEWINGTON IN 1848

Fig. 5.1 Stoke Newington in 1848.

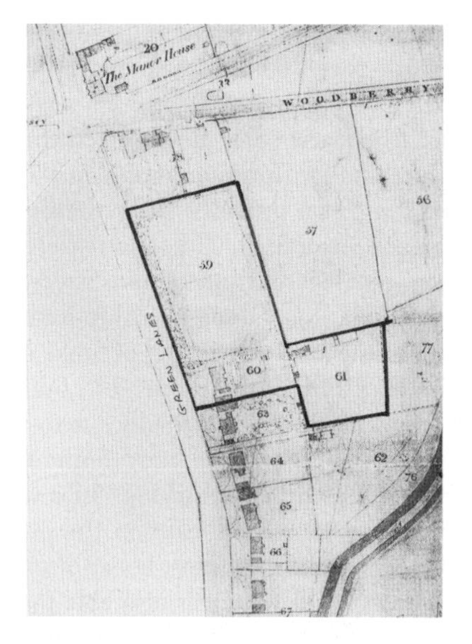

Fig. 5.2 Thomas' land leaseholding in 1841.

Fig. 5.3 Thomas Hancock in 1841, painted by T. Overton.

There is no definitive record to indicate when the house was built but there is a record of Thomas taking out an insurance policy with the Sun Insurance Company for 'Marlborough Cottage, Stoke Newington' on 7 December 1836[4] (Fig. 5.4). The house was unlikely to have existed in isolation before the northerly expansion up Green Lanes and was probably a new build at that time, since some unusual features suggest that Thomas had been involved in its design. We know[5] that his laboratory, protected by an iron door for both privacy and to protect the remainder of the house should there be an accident there, was to the left of the front door and that it had a separate staircase to his bedroom above so that, if inspiration struck in the night, he could descend without disturbing the rest of the household and make some experiments. He called the house Marlborough Cottage to distinguish it from Marlborough House to the north. Given the origins of the Hancock family this was a remarkable coincidence. In 1841, Mr Bettely, who lived in Marlborough House, possibly seeing the size of Thomas' instant family, proposed to Thomas that he should take over the meadow between them except for about half an

Fig. 5.4 Marlborough Cottage at the turn of the nineteenth and twentieth centuries (with tram lines visible in the foreground).

acre that he wanted to extend his garden.[6] This gave Thomas a further 2 acres (plot 59) and prevented any in-filling between the two properties.

Responding as chairman to a lecture presented by Messers B. D. Porritt and H. Rogers to the Institution of the Rubber Industry some 60 years after Thomas' death,[7] his great-great-great-nephew, Walter C. Hancock summed up Thomas' character thus:

> He was a man of austere character, with a very high sense of duty, and with that peculiar English trait, an absence of the capacity to show his sympathy and affection towards other people. That he was both tender hearted and affectionate was shown by two facts, namely, that he had adopted educated and provided for the nine orphaned children left behind in Cornwall by his brother John, and that although ordinary people never dared to take any liberty with him, a little child could do practically what it liked with him; a child with its peculiar innate quality of seeing through the external and appreciating beneath the kindness of the heart that beat there.

Before leaving for Cornwall the older children had attended the congregation of Joseph Burrell in Titchfield Street, and now they

Fig. 5.5 Joseph Burrell.

were welcomed back into the congregation as if they had never left, and because of the importance of their connection with Mr Burrell (Fig. 5.5), a few words on his ministry are essential.[8] When William Huntington died in 1813 his flock was scattered to the four corners of the city. His daughter, who had inherited his chapel, took a dislike to Joseph Burrell and turned him out, refusing to allow him to preach there despite all the deacons pleading with her.

Disconsolate, he sought another base from which he could minister to those of Huntington's flock who wished to remain with him. A place was found in a nearby mews and on 8 August 1813 he gave his inaugural address to a crowded congregation. He spoke with great nervousness of his own shortcomings, his personal divine revelation, and how it had led him, just as it had Huntington, to embark on a life of preaching to any who would hear him.

In truth, his early preaching days did not inspire many of Huntington's followers and most left in the months following, including many of the most influential members. This decline in his fortunes led him to take rooms at 9 Great Titchfield Street, over a room licensed for preaching, from which he also dispensed patent medicines. Here, a small company of his faithful and steadfast friends met, becoming fully united in spirit, both with him as their pastor and with each other.

Every Monday evening, a company assembled at his house with whom he used to converse in a most beautiful and edifying manner of the things that belong to salvation, being sometimes led out in a remarkable manner to dwell upon the great truths of the Gospel and the work of the spirit of God, and that those who heard him could not withstand the blessed influence and felt that faith came by hearing. The joy of his heart used to shine in his countenance and the love and tenderness of his spirit was sweet indeed.[9]

Thomas Hancock was one of those who kept faith with him and the coincidence of the arrival of John's nine children needing a tutor together with the decline in Mr Burrell's fortunes became a most fortunate circumstance for them both. This silver-haired gentle old man who had been raised in the best society in France and Prussia in his youth and with a first-class education, which he had further embellished with studies in art and music, became their mentor and guide. Through him they learned the refinements that many a society lady might have envied and through him they found the spiritual comfort that as orphans they most certainly sought.[10] It was the perfect answer, as it left Thomas free to pursue his business.

His new partnership with Charles Macintosh & Co. brought with it further challenges, beginning with the first of many patent cases in which he was to become directly involved. As long ago as 1824, Macintosh had approached Wynne Ellis, perhaps the foremost silk merchant in Britain, for financial backing. Ellis sent some of his fine silks to Macintosh for treatment in Glasgow but was not sufficiently impressed with the results to induce him to invest in the process. Now, in 1836, Macintosh and Hancock were advised that Everington and Ellis were advertising 'Fanshawes Improved India Rubber Cloth', which seemed similar in all respects to that which they manufactured. They had no option but to proceed in the courts. While they were preparing to defend Macintosh's patent of 1823, before it expired two years hence, they applied for an extension. This the court refused to consider until the following year, when the present case would have been decided. There was a great deal at stake now, for if they were to lose, their business would have no protection from unlimited competition and their domination of the market would be completely swept aside.

The case was eventually heard in the Court of Common Pleas in February 1836 before the Lord Chief Justice, Sir Nicholas Tindal, and a special jury.[11] The outcome was of such importance that the plaintiffs, Hancock and Macintosh, were represented by the Attorney General himself. His masterly address to the court so perfectly captured not only the particulars of this case but the generalities of English Patent law that it is here given in full, just as Macintosh and Hancock heard it:

My Lord and gentlemen of the jury My learned friend Mr Watson, has stated the various issues formed and you, gentlemen of the jury, will have to decide whether Mr Macintosh's patent for making waterproof cloaks and other articles, is valid or invalid. The cloaks are now very generally known, and have obtained great celebrity, and are of the greatest utility; so much so that the patent has become almost as well known as the celebrated Mr Watt's for steam engines. Gentlemen, the patent was taken out by Mr Macintosh in 1823 and has been respected generally by the public, till Messrs Everington, Wynne Ellis and others acting in concert with them, within a few months have infringed the patent right. The circumstance of no one having before called into question the right of the patentee, till within two years of its expiring, may be taken as a strong presumption in favour of the validity of the patent. Now, for the first time, is the invention invaded; and an attempt is made to upset the patent, by stating that Mr Macintosh was not the inventor; that he had not fully specified the nature of the process, sufficient to enable a workman to pursue the invention, as required by the conditions of the patent. Before the patent, many endeavours were made to obtain a substance, which in rendering fabrics waterproof or air proof, should still retain flexibility. There were cements of different kinds attempted to be used, but without success. Mr Macintosh applied India Rubber or caoutchouc, I shall call it India Rubber. The cement is made by dissolving that substance and that substance alone; the great object was to obtain a solvent for India rubber, which solvent should be evaporated; and thus leave the India rubber as a cement between the two fabrics, thereby making those fabrics impervious to air and water. The materials so combined became of great value, being flexible as well as air and water tight. He specified his patent in the following manner:

Now whereas caoutchouc in a state of solution and dissolved in the manner hereinbefore described is well known to chemists and others, and not new, therefore I do not claim any right title or privilege in respect of the same; but a manufacture of two or more pieces of linen, woollen cotton silk leather or paper or other the like substances, any or either or any combinations of any or either the same, cemented together by means of flexible cement, in manner hereinbefore described, being to the best of my knowledge and belief, entirely new, and never before used in these kingdoms, I do hereby declare this to be my specification...

Thus gentlemen, you will perceive that Mr Macintosh fully and fairly describes the nature of his invention; he describes the process by which he carries the same into effect; he points out what was before known, and distinguishes that of which he claims as his invention, which consists in applying of the flexible cement, composed of dissolved India rubber, in the manner he described, by putting it between two surfaces of fabric, and then pressing them together, so that they should be united and become, from the properties of the cement, air and water tight. The patentee had considerable difficulty to encounter in bringing this invention into general use. There was prejudice got abroad, that although the cloaks kept out the rain, they stopped perspiration, and hence were injurious to health; but after going to great expense, the use of the material became popular, and the invention has only lately become very profitable. There was, some time back, negotiations between Mr Macintosh and the defendants for entering into partnership, but it went off; and now the defendants were making cloaks, and other articles, precisely according to the specification I have read to you; that is to say, according to the substance of it, and certainly upon the principle on which the specification was founded. It therefore became necessary that the plaintiffs should vindicate their property, and hence through this action various pleas have been put on record. They say no patent was granted; in answer to this, we have the patent in court. They also say that they have not infringed. I shall be enabled to prove most distinctly that they have. I shall show that, in a most public manner, they exposed these descriptions of articles in their shop for sale. We have some which we purchased. I am, however, inclined to think that the two pleas on which they will principally rely, are, firstly that this process was publicly practised before the granting of the plaintiff's patent; secondly, there is no specification. I propose to make a few observations on each of these pleas. With respect to the novelty of the invention, there had been before this patent many attempts on this subject, and various experiments; that is the case with every invention; and if it were to be decided that a patent is invalid because experiments have been before made which approach the patented invention, which have come even within a step of it, were such held to be the law, no patent could possibly be established. Many of the most important inventions by which the manufactures of this country have been so greatly improved, have been but one step beyond what has been done for ages, and many have been but one stage beyond fruitless experiments which have been made and abandoned. Various cements have been made, but they, not being flexible, did not meet the object desired; there were in all those instances other ingredients used, which injured the property of the other material or India rubber; therefore, all previous trials were unlike that of Macintosh, and this is proved by the fact that, until his patent, such waterproof articles were not to be purchased; indeed they were not known. I will now turn to the specification. I understand from what has taken place in the court of chancery, that my learned friend will endeavour to show that the solvent used by the patentee is not properly described; it is called coal oil, which is one of the products in the distillation of coal for making gas. Gentlemen, in considering this specification, we do not undertake to show how coal oil is manufactured; we state we use coal oil, and that being a material that is known, and can be purchased, that is sufficient. It is not to be expected that an inventor of one thing shall of necessity know how to make all the materials used in the production of his invention; it is sufficient that such materials should be known and purchaseable [sic]. The excellence of coal oil is this: it is highly volatile, it having dissolved the India rubber, is quickly evaporated, and leaves the India rubber between the two surfaces of cloth, by which the fabric produced is rendered flexible and waterproof. I will now call before you the evidence by which I shall be able to show that the invention is highly useful, that it may readily be carried into effect from the specification, and also that the defendants have infringed the patent.

The case for the defendants was built on four pillars: first, that the patent did not exist, second, that double textured garments had been produced in Demerara since the end of the eighteenth century using latex as the adhesive, third, that Charles Green had used rubber solution and the double texture procedure to manufacture balloons, and finally that it was obvious by inspection of the current output from the Macintosh factory that it bore little resemblance to that produced in 1823 and therefore the process must be different and the patent could not apply. One by one these points were demolished by the Attorney General. The patent *did* exist and he produced it in court. The point concerning Charles Green was overcome when it was shown that his double textures were merely overlapping seams using rubber as a mastic while the final point was irrelevant as it did not matter how Charles Macintosh & Co. were making double-textured fabrics in 1836 provided that the method in the original patent actually worked and was being unlawfully copied by Messrs Everington and Ellis. So far so good, but now came the real test.

It could not be contested that in 1798, S. G. de La Vega had made double-skinned containers using latex to bond chamois leather skins together to provide strong leak-proof bags which could be used to convey mercury from South America to Europe.[12] Surprisingly perhaps, since the plaintiffs had emphasized that the solution was not important but that the patent referred to the double texture, the defendants did not force the issue on this point and it was agreed that what happened using latex was not the same as when using rubber solution.

The case droned on for three days, at the end of which the Lord Chief Justice began his summing up but was interrupted by the jury who declared they were satisfied and that their mind was made up to return a verdict in favour of the plaintiffs.

While Hancock and Macintosh were mightily relieved at the verdict, their victory was not complete, for opposition to the renewal of the 1823 patent became so strong that they thought it wise to withdraw their application. This left them with only one way of protecting their position; they must finally patent the masticator and the spreading machinery. Now the secret of how Thomas Hancock had been processing rubber was out for all to see and how surprised they must have been when they looked for the first time at the drawings of his 'pickle', as they first appeared in his 1837 patent 'Dough Waterproofing' (Fig. 5.6).

Lady Luck however dispenses good and evil fortune with equal impartiality. No sooner had the dust settled on the patent case than disaster visited the Macintosh factory. Thomas Hancock recalled:

In the summer of 1838 we were visited by a great calamity. Our works at Manchester took fire at midnight, after all the men had left, and although the building was constructed on what is called the fireproof principle, such was the inflammable nature of the large amount of stock it contained, that all that was combustible was speedily destroyed, and what was worse, several men lost their lives by the unexpected falling of some of the heavy machinery through the arched floors, supposed to have been secure against such a casualty. I happened at the time to be in Scotland, on a visit to my late much esteemed and lamented friend and partner Mr Charles Macintosh. I hastened to Manchester, and soon witnessed the scene of devastation, and heard of the appalling circumstances in the melancholy loss of life of the poor men who had ventured to stay too long in the devoted mill. The walls were standing, and the chimney and stone stairs, and most of the arched floors, but the machinery

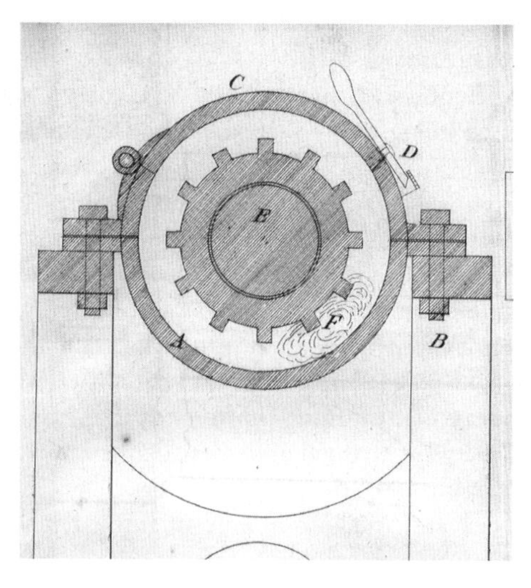

Fig. 5.6 The Iron Masticator illustrated in Thomas' patent of 1837.

was damaged beyond repair, and the reservoirs of solutions and solvents being chiefly at that time of wood, had of course disappeared. We never could ascertain the cause of this disaster: it remains a mystery to this day. Having plenty of zealous help at hand, the rubbish was soon cleared away; in the meanwhile no time was lost in devising means for resuming work. This was the more pressing as our business was yearly on the increase, and the season for the demand of our goods was fast approaching.

Despite this huge interruption to their production, Thomas was able to report[13] that the firm did more business in that year than in the previous one, turning out between 3,000 and 4,000 square yards of double textured waterproofed cloth each day.

Meanwhile, back in Stratford, Walter was able to indulge himself by running his largest steam carriage yet just for the pleasure of it.[14] He had originally been persuaded to build a more powerful steam drag for a customer who then reneged on his contract, so he converted this machine into the largest carriage he was ever to build. In view of its large size and power he named it 'Automaton' (Fig. 5.7).

Writing to the editor of the *Mechanics Magazine* in September 1836, he gave an account of its performance:

Since the last notice in your magazine a new carriage, the 'Automaton', has been brought upon the road, the only difference between which and those

Fig. 5.7 Automaton.

preceding it is, that the engines are of greater power (having cylinders of twelve inches diameter, whilst those of the others are nine inches) and the carriage of altogether larger dimensions than the others, it having seats for twenty two whilst they are only calculated for fourteen passengers. It is an open carriage like 'The Infant'; and although only calculated for the accommodation of twenty two passengers, it has carried thirty at one time, and would then have surplus power to draw an omnibus or other carriage containing eighteen more passengers, without any material diminution of speed. Its general rate of travelling is from twelve to fifteen miles per hour. On one occasion it performed (when put upon the top of its speed and loaded with twenty full grown persons) a mile on the Bow Road, at the rate of twenty one miles per hour....

And he went on to summarize the results of his 12 years' experience remarking:

I cannot conclude without noticing, with gratitude, the general civility and attention which I have met with, and my pleasure in discovering that the antipathies which existed in the earlier part of my career are gradually subsiding, and that, in fact, never now meet with incivility excepting with a few carters or draymen, who consider the introduction of steam carriages as an infringement upon the old established use of horse flesh.

Years of practice have now put all doubts of the economy, safety, and superiority of steam travelling on common roads at rest, when compared with horse travelling; and I have now in preparation calculations founded upon actual practice, which, when published will prove that steam locomotion on common roads is not unworthy of the attention of the capitalist, though the reverse has been disseminated rather widely of late by parties who do not desire this branch of improvement should prosper against the interests of themselves.

Mr Hancock, remarked the editor upon this letter

is now the only engineer with a steam carriage on any road. Sir Charles Dance, Colonel Maceroni, Dr Church, Messrs Ogle, Summers, Squire, Russell, Redmund, Heaton, Maudsley, Frazer, and a host of others—where are they? Echo answers 'where'? Strange to say however, we see steam carriage companies advertised, whose engineers have either never yet built a carriage, or whose carriages when built, have never stirred out of the factory yard.

Walter Hancock drew his *Narrative* to an end with the following words:

The writer has only to observe, in conclusion, that much injury has been done to the cause of steam locomotion on common roads, by abortive and injudicious attempts to gain momentary reputation, by the acquirement of a rate of speed not economically attainable in the present state of the invention; and while some, after an immense cost to the parties engaged, have either from difficulties they had not foreseen, or duly considered, or from impediments of a local nature, been compelled to abandon their undertakings, others, having reached a certain point, and unable to get further, have had recourse to meretricious display, floundering along at great speed, until exhaustion, at the end of two or three miles, brought them to a stand. The consequence has been a certain distrust in the public mind respecting the feasibility of the subject, which has required, and may still require, much time, and a series of convincing experiments, fully to dissipate.

The writer believes that at this moment he has not one competitor who, singly and steadily pursuing his course, has borne up against all opposition, and fought the battle to such a successful issue. In commencing the undertaking, he was fully aware that nothing short of a toilsome and steady perseverance could accomplish his purposes, and produce a steam carriage of real utility in the public service.

A drive round Hyde Park, or any other well-made level road, was a matter soon and easily achieved; but when the ordinary turnpike roads are attempted,—good and bad as we find them, rough and smooth, hard and soft, hill and dale, long and deep reaches of new laid metal—then and not until then, the decisive trial commences.

In such a course of experiments, the Infant took the lead, and plied for hire on the Stratford road, and Mr Gurney shortly afterwards followed at Cheltenham. The latter road, and the centre or gravelled portions of the former, were much alike, but the Stratford road has stone paved portions on both sides, and on these paved parts, not always in the best condition, The Infant was driven; and this often by choice, for the sake of experiment and observation; the writer thus obtaining some of his best, although most expensive, lessons.

Setting aside therefore all exaggeration, the writer feels he has now demonstrated practically, that steam travelling at the rate of from ten to fourteen miles an hour with passengers, and for goods and merchandise at from five to seven miles an hour, can be effected safer, and on a scale of charge much below the present horse conveyances, and with a far greater return for capital employed.

All this was written in 1836, when the experimental period was at an end, but in 1838 Walter Hancock added a post-script to his *Narrative*:

The preceding sheets were printed nearly two years ago; since that time I have brought out the Steam Phaeton [Fig. 5.8]... intended for my private use; it has seats for three persons, independent of the one steering. It has run principally in the City and upon the roads in the east of London; but, within the last few days, I have occasionally run it in several parts of the west end of the town principally in Hyde Park, amongst the throng of carriages which are always to be found there on fine afternoons at this period of the year. Of course it did

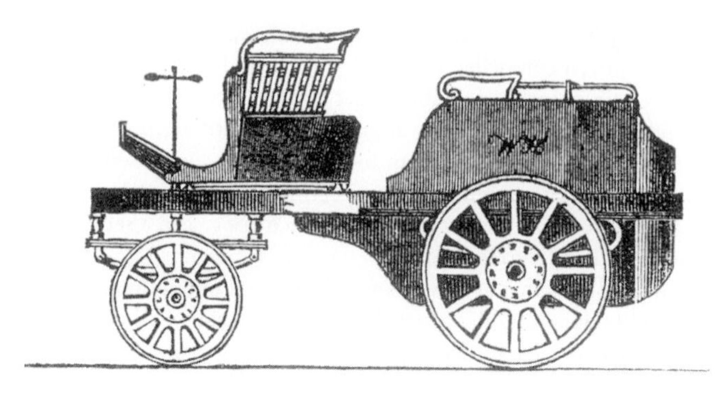

Fig. 5.8 Walter's Steam Phaeton.

not fail of attracting notice, and as there was no noise, nor any appearance of steam, fire or smoke, I was gratified to witness the general expression of approbation, as well as particular enquiries of several noblemen and gentlemen, some of whom were pleased to request a ride with me.

I have, with this carriage, gone at the rate of twenty miles an hour, but its more usual rate is not more than from ten to twelve. My object in building it was to demonstrate that my boiler is applicable to the propulsion of carriages for actual use on common roads, of any or every degree of power; I am now engaged in applying this boiler to a locomotive engine for a railway train, from which I confidently anticipate very considerable advantages to arise.

However, Automaton still had a contribution to make to the story of steam carriages. The *Observer* of 8 October 1839 reported that:

On Monday last Mr Hancock's steam carriage, The Automaton, made a journey from London to Cambridge in four and a half hours after leaving the metropolis.

A more detailed look into the journey[15] shows that this was its first journey for some two years, during which time it had been exposed to the elements in Walter's yard. It made a stop for repairs of just over one hour in Cheshunt and then stopped at The Oak in Ware for lunch—which occupied a surprisingly brief 50 minutes! Crowds gathered to watch it tackle Wade's Mill Hill, a two mile steep gradient 'with a soft bottom and newly covered with loose gravel' and were astonished when it ascended with no problems—more than can be said of some motor cars 30 or so years ago in wintry conditions! That leaves something over two hours driving time, which is probably about the same as it would take today if one used Walter's route, the A10!

In Cambridge Walter met up with his brother, Charles, who was by this time considered one of the outstanding animal painters of the period and who was *en route* to the racing at Newmarket, where, no doubt, he hoped to meet some potential clients in his position as resident artist at Tattersalls. It was probably a very relaxed meeting at the 'University Arms' and the next morning, Tuesday, Walter had his one and only crash as he attempted to take a corner too fast and ran off the road.[16] The carriage was soon repaired and on the Wednesday journeyed to the racecourse but it was not beyond one wit to wonder if this was the first recorded case of 'driving under the influence'.

Walter made only one more journey in the Automaton before his preoccupation with commercial steam carriages came to an end. On 20 March 1840 he drove from his factory in Stratford, via Finsbury Square, where he picked up a party of gentlemen, to the 'Green Man' in Barnet and home again 'after taking refreshment'.[17] The journey was uneventful, but there is no record of the refreshments of which Walter partook!

So concluded Walter's involvement with steam on common roads but for a short time he endeavoured to follow the market by designing equipment for railways. In April 1838 he issued a 'PROSPECTUS OF HANCOCK'S PATENT SAFETY STEAM BOILER AND LOCOMOTIVE ENGINE MANUFACTURING COMPANY FOR THE MANUFACTURE OF PATENT SAFETY BOILERS applicable to steam vessels and steam carriages of all kinds and improved locomotive engines for railways and common roads, secured by Letters Patent'.[18] In it he made various claims including the fact that John Farey, one of the most eminent engineers of the day, bore testimony to their superiority over all others before a Committee of the House of Commons in 1829, and many improvements had been made since that date. Unfortunately for him, technology had already progressed beyond his rather narrow specializations and insufficient shareholders came forward to get the project launched. He did build a locomotive for the Eastern Counties railway and he patented a new railway signalling system, as well as designing and patenting a feathering paddlewheel for steam ships, but in April 1842 he gave up his yard and factory beside Stratford High Road, and the lease was sold by public auction.

The sad truth was that he had run out money, and could no longer afford the rent. Machinery for Charles Macintosh & Co was now purchased in Manchester, and the excellent designs that Walter had conceived had proved so efficient that no new designs were required. Orders for valves for the inflatable goods made at Goswell Mews had

56 STRATFORD.

A HIGHLY VALUABLE AND IMPROVABLE
FREEHOLD ESTATE,
Extensive range of Buildings and Land,
Producing £60 per annum,
TO BE SOLD BY AUCTION,
By Mr. T. Harvey,
On Thursday, April 21st, 1842, at Twelve o'Clock, at the Auction Mart, near the Bank, London, by Order of the Proprietor, and with the consent of the Mortgagee, in One Lot,

COMPRISING

ALL that Substantial Brick-built RANGE OF BUILDINGS, consisting of Coach-houses, Stabling, with Lofts, Granary, Warehouses, Sheds, Workshops, Yards, Premises, and small Pasture Field, with an extensive frontage, abutting upon the great high-road, in the centre of the populous town of Stratford, Essex, as the same are now in the occupation of Mr. Hancock, engineer, and used for the manufactory of his patent steam carriages; Mr. Woodcock, and others, producing at low rents £60 per annum.

The Property is let to the above tenants from year to year, and from its superior situation, frontage, and other advantages, it is presumed it may be very greatly improved, and is truly desirable for a first rate manufactory or building speculation.

May be viewed by leave of the tenants any day, and Particulars with Conditions had ten days preceding the Sale; at the Inns at Stratford; at the Auction Mart, London; of Mr. Harvey Jones, solicitor, Austin Friars; and at the Offices of Mr. T. Harvey, Auctioneer, &c. Ilford and Romford, Essex.

Fig. 5.9 Public auction notice for Walter's factory and yard.

been falling off for some years and were now insufficient to pay for the yard and workforce. The ignominious task of reducing the steam carriages to their component parts was forced on him so the parts could be sold. Walter may have hoped that Thomas would throw him a lifeline, but this did not happen for reasons that will soon become clear.

With no current income, and faced with debts he could not pay there was only one end to this chapter of his life and that was bankruptcy. On 27 December 1843 he filed his petition in the Court of Bankruptcy, where he was summoned to appear for examination on 13 January 1844. He was declared a bankrupt on 3 February 1844. At the creditors' sale the only assets they could find were his patents, but history does not record whether there were any bidders (Fig. 5.9).

Why did his brother not throw him a lifeline, when he could well afford it? Thomas had already supported Walter over the years to the tune of £15,000[19] and had thought this a sensible arrangement, as it gave him free design and assembly facilities in the Stratford yard and there was always the possibility that steam on the common roads might have become established and profitable. Now that there was no longer any chance of this becoming a reality, any further financial assistance which Thomas might have provided would merely have put his money straight into the hands of the creditors.

For Walter bankruptcy was not only a depressing humiliation, but the end of the dream that had sustained him for the last 20 years. It had all ended in abject failure. He seemed at the end of the road, and, ominously, he began to suffer from chest pains and gout.

6

Life's Ups and Downs

Some degree of retrenchment needed for Thomas and Macintosh as business slows—J. L. Hancock buys what was John's business from Macintosh and takes over the Goswell Mews factory. Thomas sees vulcanized rubber for the first time—Nathaniel Hayward and Charles Goodyear—their history, relationship and the discovery of vulcanization—Goodyear's lack of scientific knowledge and business acumen—attempt to sell (inadequately developed) process to Macintosh—Thomas sits back and awaits developments—none forthcoming—Macintosh dies—his son examines his character. Thomas realizes that Goodyear has no patentable process so feels justified in doing some research of his own—rapid success and soon fully understands the process—patents it a few weeks before Goodyear finally attempts to do so. Question—who discovered vulcanization?

One of the reasons Thomas and Charles Macintosh decided not to continue with their application to extend Charles' 1823 patent was the fall-off in demand for the (almost) eponymous mackintosh. For many years they had contended with the disapproval of the medical profession, which argued that close-fitting rubberized fabrics were injurious to the health of the wearer.[1] In fact there was no disagreement since the partners had always advocated loose-fitting capes and mackintoshes, but fashion, and recalcitrant tailors, preferred more tightly fitting garments. More importantly and something about which they could do nothing was, ironically, the rise of steam locomotion. As more and more travellers moved from the outside of coaches to the inside comfort of railway carriages, it was inevitable that demand for the waterproof cape of the coachman and the mackintosh for the traveller would decline.

Thomas' ever-expanding product portfolio certainly cushioned this lowering of demand. As he wrote:

Although the waterproof trade had somewhat declined, yet we still did a considerable business in that line, and the air-proof department, I think, never flagged much. Life-preservers became more known, and their utility had been proved in the saving of life. The demand for diving-dresses and the tubing for them increased as well as the trade in hose-pipe, fishing-boots, &c., and the trade continued good for fine cut sheet-rubber, and blocks for stationers, &c. The trade in solutions had now become considerable; not only for rendering the seams of garments waterproof, and for shoemakers, but as a cement also, which for some purposes it is particularly well adapted, possessing this peculiarity—that it remains adhesive when dry, and two surfaces coated with it, when brought into contact, immediately unite, requiring only a little pressure or rubbing with the hand. This cement will be used extensively when more known.

The long-sightedness of choosing to patent the masticator and spreading table was very apparent!

Nevertheless, it was decided between Thomas and Macintosh that some retrenchment would be sensible. This took the form of a retreat from London and the concentration of all activities in Manchester. Some years earlier the son of Thomas' eldest brother, James, also called James but with the addition of his mother's family name—Lyne—to distinguish him from his father, had joined Thomas' business and had been an excellent student, rising to the position of Thomas' chief assistant. He believed that the business had much to offer and so, probably with financial backing from Thomas, he purchased the plant and stock and arranged to lease the Goswell Mews factory from Charles Macintosh & Co. He also agreed to act as agent for some aspects of the Macintosh business in London in exchange for the contracts for hose pipes and Deckle Straps with which, in future, Charles Macintosh & Co. was not to interfere. This was to lead to some heated arguments and correspondence between the various bodies and individuals in a few years time, but in order to maintain a degree of chronology there is another matter to deal with first.

In his *Narrative* Thomas simply wrote:

Some time in the early part of the autumn of 1842, Mr. Brockedon showed me some small bits of rubber that he told me had been brought by a person from America, who represented himself as the agent of the inventor: it was said that they would not stiffen by cold, and were not much affected by solvents, heat, or oils. Mr. Brockedon told me that the mode of manufacturing this rubber was a secret, and that the agent who had shown them to him, declared himself totally ignorant of it.

If Thomas had been surprised to discover Macintosh's patent of 1823 he must have been amazed by these small pieces of rubber. As a deeply religious man, he could not describe them as the Holy Grail of rubber product manufacturing but he must have searched hard in his mind for an appropriate description! In truth, although he had mixed many chemicals with his masticated rubber, he had never expressed any specific intention or desire significantly to alter the properties of the raw rubber which he had used for over 20 years; his concerns had been in its processing and in using it to make products suited to the properties of which he was aware. Now that he had seen the extended properties of this treated material, he saw a new world opening up before him; but how had this transformation come about?

To answer that question we have to cross the Atlantic to the east coast of America, for it was here that the nineteenth-century industrial expansion began and it was no more evident than in the areas around Massachusetts and Connecticut. Here, in the 1830s, we find reference to Nathaniel Hayward (Fig. 6.1), an illiterate foreman in the factory of the Eagle Rubber Company.[2] Many hopeful firms experimenting with rubber had been bankrupted in this decade, and this one was no different. By 1838 it was in considerable difficulties and Hayward took it over himself. His interest lay in treating rubber sheets so that the surfaces lost their stickiness, and he discovered that if he rubbed sulphur into the surface and exposed the sheets to the hot sunlight the stickiness disappeared. This was a totally contra-intuitive process since it was well known that exposure to sunlight was one sure way of bringing about the degradation of any rubber product. Hayward called his process *solarization* and it can be argued that, as is written in the preface to the facsimile reproduction of Charles Goodyear's autobiography, *Gum-Elastic,* he was *the true discoverer of* vulcanization,[3] although this denies credit to Dr Freidrich Ludersdorf who published similar findings a few years earlier.[4] Certainly, Hayward would not have come across that work.

Charles Goodyear had been born in 1800 in New Haven, Connecticut (Fig. 6.2). He entered the hardware business with his father, but the venture failed in 1830 and he started looking for something to interest him. In early 1834 he came across rubber indirectly when he saw in the store of the Roxbury India Rubber Co., America's first rubber manufacturer, some rubber life preservers, or life jackets as they are called today.[5] His interest was in the inflating valve, which he thought crude and potentially dangerous so he designed one he considered to be an

Fig. 6.1 Nathaniel Hayward.

improvement and took it back to the store. The manager was not interested and showed Goodyear racks of rubber goods which had melted to a stinking gum in the heat. Goodyear took his first good look at rubber and proclaimed that:

There is probably no other inert substance which so excites the mind.

In the more temperate climate of the United Kingdom the rapid liquefaction of rubber in heat and strong sunlight was a problem which could, as Thomas had showed, be avoided. But this was not the case in Goodyear's home region, and dealing with this problem became his overriding obsession to the detriment of himself, his long-suffering family and his friends who had to support them, at least until their patience ran out, since he had no time to earn money while he pursued his dream of preventing rubber degradation.

In 1838, after numerous failed attempts to 'stop the rot' he met Hayward and, on seeing what his 'sulphur and Sun' treatment had brought about, suggested he patent it. This he did (USP1090) in November 1838, and Goodyear then purchased all rights under the patent.[6] Goodyear, sulphur, and rubber had joined forces. There are numerous stories[7] describing how Goodyear discovered for himself how to vulcanize rubber but even the (true?) version, which he gave in his autobiography, is so short that it leaves many unanswered questions.

Fig. 6.2 Charles Goodyear.

Here we can just accept that one day, by accident, some of his mix fell onto a hot stove. When he scraped it off, he found that it had charred but around the charred area was a flexible material, which he called 'gum elastic' and which today we call vulcanized rubber.

Unfortunately, he was unable to advance much from there! He was neither a scientist nor an engineer, his mental processes were not attuned to developing a structured investigation into the process which he had stumbled upon, and by this time most of his friends and relatives had abandoned him, refusing to advance him money to continue his work. He continued to investigate the phenomenon as best he could and occasionally produced some small, vulcanized pieces but nothing of immediate commercial value. It was at this point, with no hope of funding in the United States, that his thoughts turned to Europe. He found an Englishman, Stephen Moulton,[8] and through him he sent samples of his heat-and-sulphur-treated gum to England in 1842, where he asked Thomas' friend William Brockedon to give him any introduction he could to the trade. Thomas records that:

> Mr. Brockedon gave him the names of Macintosh and Co., and the representatives of the Caoutchouc Company, (and, I think, at least one other) with whom the party also left specimens. Macintosh and Co. told the agent that as he could give no information, they could not judge of the merits of the invention, as it might be easy to make these small specimens, whilst difficulties

might be found in its application upon a large scale; nor could they judge whether their present appliances, which had been very costly, would be suitable for this manufacture, or whether any modification of their present plant would answer, or if the modes were so altogether new as to require a fresh outlay. Under these circumstances, as they could not act in the dark, they recommended the taking out of a patent,[9] when a clear comprehension of the whole could be obtained, and the invention openly dealt with according to its merits. This course was approved of, and the agent said he would immediately advise his principal to take out a patent.

Goodyear had demanded £50,000[10] for his secret, and it must have been fascinating to have been a fly on the wall of the boardroom when the partners were discussing the options. No doubt the words *pig* and *poke* flew across the table and possibly some cynic went as far as to shout *snake oil*, but logic and business practice ruled the day. If Goodyear could patent his process then they would be very interested in looking at its value to the company. Goodyear's problem, however, unknown to all the English, was that he could not patent the process because, as was noted earlier in the argument between Thomas, Macintosh, and Ellis, etc., the Attorney-General described the fundamental principle of a patent to be 'sufficient to enable a workman to pursue the invention, as required by the conditions of the patent'. All Goodyear could offer was rubber, sulphur, lead (not known to Thomas), and heat. It was not enough.

Thomas was a man of honour and principle so he put the samples to one side and waited for the patent to arrive. He had realized that sulphur was involved somehow, as he had used it himself in some early experiments and could recognize its bloom on the surface of the pieces. He confirmed that the treated rubber was resistant to the solvents one might expect to dissolve it and that it did indeed maintain its flexibility and stability over a wide range of temperatures. The patent would be worth waiting for! To his surprise it never came, and in spite of a further request through Stephen Moulton that the process be patented so that Charles Macintosh & Co. could consider it, no more was heard from the United States.

Thomas had other things to think about. For a number of years Charles Macintosh had been in ill-health, and on 17 July his condition deteriorated significantly. Word was sent to Thomas that Charles wished particularly to see him and he set out immediately for Glasgow, but he was too late. Charles Macintosh died on 25 July 1843. Thomas noted in his *Narrative*:

I have never met with any person for whom I entertained a greater esteem, nor could any two persons in business act more cordially nor with more frankness, and I believe mutual esteem, than subsisted between us.... An interesting biographical memoir of him, written by his son and successor, the late Mr. George Macintosh, was printed for private circulation, but was never published.

The relationship Thomas described as existing between himself and Charles Macintosh is easy to understand, for in that biography Charles' son wrote:

In even so imperfect a sketch as the present, the omission of allusion to the religious sentiments held by the subject of this memoir, might probably be considered as scarcely pardonable.... With those persons who had marked Mr. Macintosh's course through life, with those at all acquainted with his sentiments and opinions in general, or with those who had witnessed his deathbed and death, not a doubt could exist as to the purity and intensity of his feelings of Christianity. His, however, was a Christianity freed and refined from the priestly bigotry and intolerance, and the secular superstitions and inconsistencies of but too many professing Christians, whether Catholics or Protestants; and seemed to be comprised in the divine precepts contained in the following passages of 'holy writ': 'Jesus said unto him, Thou shalt love the Lord thy God with all thy heart, and with all thy soul, and with all thy mind. This is the first and great commandment. And the second is like unto it, Thou shalt love thy neighbour as thyself. On these two commandments hang all the law and the prophets.' St. Matthew's Gospel, chapter xxii. verses 37–40.

It must also add to our recognition of Thomas' own character that he spoke so neutrally about George's memoir, since he could quite rightly have felt aggrieved that his name never appeared therein nor was his contribution to the success of Charles' business even referenced. Perhaps the most interesting part of the memoir is the collection of 57 facsimile autographs of famous people who had been in correspondence with Charles throughout his life; again without that of Thomas.

After the funeral Thomas turned his thoughts back to the scraps of vulcanized rubber from the United States. He confirmed that no patent had arrived at the offices of Charles Macintosh & Co. and wondered what to do. He was in an awkward position. He certainly had no intention of stealing another person's secret; that would be against all the beliefs round which he had based his life; so what should he do? Then it struck him; surely if he were going to impress someone or some company with an invention, then he would certainly try to show them the very best samples he could prepare and perhaps some products which could be tested for performance and longevity. So what were Goodyear and Moulton up to? The answer was blindingly obvious. These *were* the best samples which Goodyear could make, and he did not have a viable process to patent or sell. Further, since he had ignored requests for the process to be patented, he could not have advanced his process in the intervening months.

In his *Narrative* Thomas describes his course of action once that realization had dawned on him. He began by registering, albeit indirectly, his disappointment that some people had later claimed that he had somehow 'stolen' the invention:

I made no analysis of these little bits, nor did I procure, either directly or indirectly, any analysis of them. In making my experiments, I depended entirely and solely on my own exertions, having some confidence in an experience of upwards of twenty years of unceasing application to the manipulation with my own hands of the substance I was dealing with; and it having been shown that rubber could be so made as not to stiffen by cold, I devoted myself to the discovery, if possible, of some mode by which this property could be imparted; and I considered the small specimens given to me simply as a proof that it was practicable...I knew nothing more of the composition of the small specimens given me by Mr. Brockedon than what I or any other person might know by sight and smell.

There is one obvious piece of evidence to support his statement. Goodyear had included lead oxide in his mix, and if Thomas had carried out an analysis he would certainly have discovered this and included it in all his experiments. He did not!

In his *Narrative* Thomas described his approach in considerable detail, including his lack of success in many experiments:

I remember well that I had from the first a strong impression that the rubber underwent the change either when in a state of solution, or when greatly softened by heat, and these two points I kept pretty much in view. I compounded the rubber in one or other of these conditions with an almost endless round of matters of all kinds. I treated these with the rubber separately, and in innumerable combinations. In some of them I included sulphur, and employed heat in almost every case, without regarding what degree of heat I used.... When making the compounds with solvents, I first dissolved the rubber, either making the solution of a thin consistence, or of the consistence of dough, and then mixed or worked up the other matters with it. When I wished to expedite their drying, I sometimes laid them on a small metal plate, heated over a chamber lamp, or larger plates heated by the fire, and at other times I submitted them to the heat of an oven. As before mentioned, I frequently employed sulphur in the compounds, but they were not at all improved by it, not having yet by any chance used heat sufficiently high when drying them to produce the change; or if the heat was sufficient, which it most likely sometimes was, the compound became dry in too short a period to affect it (as I removed the pieces as soon as they were dry)...I therefore for a time relinquished the use of sulphur in most of the compounds as useless, and pushed on with other matters still feeling a conviction, as I think most would naturally have done, that the rubber must necessarily undergo the change in its constitution, whilst either very soft and plastic, or in a state of solution; however, after trying an endless round of mixtures in this way, with and without heat, I still failed of success. Whilst looking over some of my former experimental scraps, I saw in some of those containing sulphur, variations I could not at the time account for. Some portions were different to others in the same specimen, which for the present I could not comprehend, although I well knew afterwards...I spent all my spare time for months with these experiments; my habit was to dispatch them quickly, making them very small in bulk, throwing aside some thousands of trial scraps, and selecting and keeping for inspection any that appeared promising. During the winter months I generally found the weather cold enough to test my scraps; but as the spring and summer came on, I employed ice, purchased from an ice-cart which passed my gate every morning (from Southgate on its way to town), and many an earnest and careful examination have I made in a morning of the scraps in the ice. My experiments now had become very interesting; I had certainly produced in some of my scraps, or portions of some of them, that condition of rubber which I afterwards called the 'change'...I was ignorant of the importance of the degrees of temperature employed, as well of the like importance of the period to which any specific compound required to be submitted to heat. These points could be ascertained with exactness only by time and vigilant watching, and this I began immediately to set about; and as the law allows a patentee six months to work out his discoveries before he is called upon to enrol [sic] his specification, I applied for and obtained a patent for my inventions, which passed the great seal on the 21st November 1843.

He now had six months in which to rationalize his results and produce a full specification that could, if properly followed by a workman, provide a perfect vulcanized article. Only then would his patent stand. Now that his laboratory was in his house he could be sure of confidentiality, so coming home in the evenings he would after supper with his nieces and nephews lock himself away behind his iron door and continue with his researches. Once again using Thomas' own words:

A thought now occurred to me that in the end proved extremely valuable. Revolving in my mind some of the effects produced by the high degree of heat I had employed in making solutions of sulphur and rubber, as before stated, in oil of turpentine, it occurred to me that as the melting point of sulphur was only about 240°Fahrenheit [116°Centigrade], which I knew would not be injurious to the rubber, it would be well to see what would ensue on immersing a slip of sheet rubber in sulphur at the lowest melting point. I accordingly melted some sulphur in an iron vessel and immersed in it some slips of cut sheet rubber, about half an inch wide, and about one sixteenth of an inch thick. After they had remained some time I examined them, and found the surface had assumed a yellowish tan colour. I immersed them again; and on withdrawing them the second time, I cut one of them across with a wet knife, and found that the rubber was tinged of this tan colour to a considerable depth. I immersed them again; and on the third examination I found the tan colour had quite penetrated through the slip. This was strong evidence that the rubber had freely absorbed the sulphur, and I fully expected to find that these slips were now 'changed,' but in this I was greatly dissappointed, [sic] for on applying the tests, I found that not the least 'changing' effect had been produced. I now replaced them and raised the temperature of the sulphur, and allowed them to remain a considerable time; and on withdrawing one of them the fourth time, I found, to my great satisfaction, that it was perfectly 'changed' retaining the same tan colour throughout; the other slips remained in the sulphur whilst this examination was going on, and on withdrawing them, I found the lower end nearest the fire turning black, and becoming hard and horny, thus at once and indubitably opening to me the true source and process of producing the 'change' in all its states and conditions, and in all its pure and pristine simplicity.

By tackling the problem as he understood it in a rational and systematic way, Thomas had obtained all the information he needed. By choosing appropriate combinations of temperature and time, he could vary the level of vulcanization from light elastic to hard ebonite (see Appendix III). Realization came immediately that his masticator or mills would be perfect for introducing known small levels of sulphur into raw rubber, and so a further series of experiments added the variable of sulphur level to those of time and temperature. Without realizing it, Thomas had invented the 'design experiment' used invariably today to obtain the maximum amount of data from the minimum number of experiments when there are several variables to be investigated in one overall experimental set. The patent was duly enrolled on 21 May 1844 and was exhaustive in its scope. In an unintended touch of irony, Thomas remarked that he had also found that thin sheets of rubber could be vulcanized if sulphur was rubbed into the surface and they were then heated in steam. The wheel had turned full circle back to Nathaniel Hayward.

The process was immediately transferred from Thomas' private laboratory to the Macintosh works at Manchester, and only minimal modifications were needed to the machinery so that they could be used for

obviating the adhesiveness of manufactured rubber, not only to the surface of single textures but to various other purposes.

The process needed a more charismatic name than 'the change' and in a discussion between Thomas and a group of his friends, William

Brockedon proposed 'vulcanization' in deference to the mythological god Vulcan and his connection with heat and sulphur.

Charles Goodyear at last obtained enough results to attempt to patent his process in the United Kingdom, but he was just a few weeks too late! In fact there is evidence that his patent would not have stood even if it had been accepted before that of Thomas. Goodyear had, at last, received some financial backing from a shoemaker called Horace Cutler,[11] who wanted to make vulcanized rubber overshoes but, by the end of 1842, the two had parted acrimoniously when Goodyear had been unable to fulfil his part of the agreement. Cutler then went to work for one of Goodyear's rivals, Horace Day, their agreement being that he would pass over the vulcanization secrets that he had learned from Goodyear. The pair soon fell out, but Day had all the information he thought he needed and set out to beat Goodyear at his own game. By the end of 1843 his success rate mirrored that of Goodyear, so we can be certain that there was no viable vulcanization process in existence in the United States at that time.

Who then discovered vulcanization and changed the course of the world? A world without vulcanized rubber would be a very different place from that which we inhabit today! There can be no doubt that Nathaniel Hayward made the first vulcanized rubber, but did he *discover* it if he did not realize what he had made? Charles Goodyear did realize what Hayward had done but, only by accident did he develop a process to produce vulcanized pieces of rubber—and then only on rare occasions. Thomas Hancock was given some of Goodyear's samples and, after waiting for Goodyear's patent, which did not materialize, set about studying the reaction between rubber and sulphur in a systematic way until he fully understood how to make vulcanized articles reliably. No *one* person discovered vulcanization. Hayward did not appreciate what he had made, Goodyear only thought about sulphur because of his meeting with Hayward and was unable to advance his accidental discovery to a commercial level, and Thomas Hancock only investigated the use of sulphur because he was given some small pieces of Goodyear's vulcanized rubber. What can be claimed without doubt is that Thomas was the first to establish conditions for the reliable manufacture of a vast range of vulcanized rubber goods and has an unanswerable claim to be the creator and father of the UK rubber industry.

7

A New Industry

Applications of vulcanized rubber—first patent is for elastic bands. Macintosh's son joins the board of Charles Macintosh & Co. but leaves shortly afterwards and new board of directors formed. Thomas patents shaped moulds for numerous applications—use of ebonite (discovered during his investigations into vulcanization) for making moulds and finished products. Discovery of cold curing process by Alexander Parkes allows greater choice of dyes and colorants in final products. Thomas and Hayward agree that the latter can import rubber overshoes into the UK (for a consideration).

Thomas had created a completely new industry, although he did point out that even his old business, before the discovery of vulcanization, was using 3–4 tons of rubber per week. The uses to which vulcanized rubber could be put were yet to be realized and would continue to expand through to the present day—and almost certainly beyond. The first patent taken out for a product made of vulcanized rubber was that of the rubber manufacturing company Messers Perry and Co., Rubber Manufacturers of London, in which Mr Stephen Perry patented the humble yet ubiquitous elastic band. The patent was applied for in December 1844 and granted on 17 March 1845. Just one item among the thousands to change the world we live in! Among the numerous items which Charles Macintosh & Co. manufactured were solid rubber wheel-tires (note the American spelling) which, Thomas observed:

> are about an inch and a half wide, and one and a quarter thick. Wheels shod with them make no noise, and they greatly relieve concussion on pavements and rough roads; they have lately been patronized by her Majesty.

Thomas Hancock's own brougham, shown in Fig. 7.1, was the first carriage to be fitted with rubber tyres.[1]

Fig. 7.1 Thomas Hancock's personal brougham (being used as a summerhouse at the time of the photograph, many years after Thomas' death).

There is nothing to be gained by further describing the products which were being manufactured by this process; the reader could sit down with a pencil and paper and spend a day or more listing them. Even the several hundred rubber components in a modern motor car would be found under the generic headings of seals, gaskets, and buffers, while the pneumatic tyre was first invented in 1845 by Robert W. Thompson, a Scottish engineer. His elegant patent involved a series of inner tubes similar to those used in bicycles today and individually inflated with air before being encased inside a heavy rubber or leather outer tyre stretched around the rim (Fig. 7.2). Thompson's design was an improvement over the solid rubber tyres, but it was not a commercial success since it preceded the existence of both the bicycle and the automobile. It was left to John Boyd Dunlop to take the credit over 40 years later, although his application for a patent was rejected on the grounds of prior knowledge.

With the death of Charles Macintosh, his son, George, joined the board, and on 31 July 1845 another of the founding partners who was by now one of Thomas' old friends, Hugh Hornby Birley, died. Thomas described him thus:

> He was a most estimable man in all the relations of life, and with uprightness of character, he possessed a calm and dignified amenity, which endears his memory doubtless to many, and especially to me.

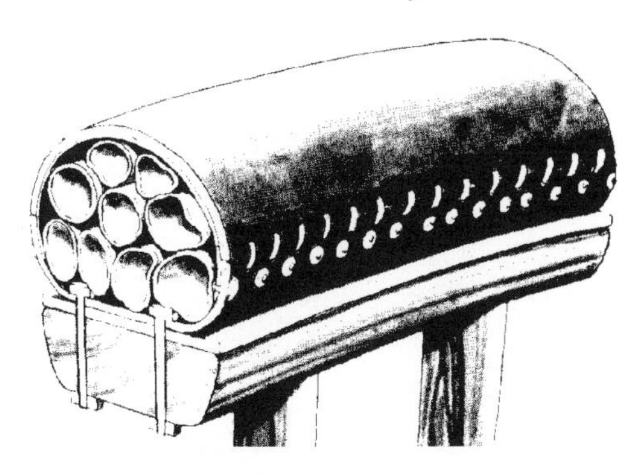

Fig. 7.2 R. W. Thompson's first pneumatic tyre.

It was time for a change at Charles Macintosh & Co. and on 1 November 1845 a Notice of Dissolution of Partnership[2] was drawn up showing that another of the Birley brothers, Henry, had retired together with George Macintosh (Fig. 7.3). The remaining partners were, in order of signature, Richard, Thomas and Herbert Birley, Thomas Hancock, and William Brockedon: three with a cotton-weaving background and two with a knowledge of rubber. George's rapid departure from the board after just two years is not explained. His lack of interest in matters relating to rubber can be gathered from the small number of pages devoted to it in his father's biography and his completely ignoring Thomas' contribution to the firm's development. This could indicate some animosity, but there is no record, or even hint, of it in either of their writings.

Thomas had no time to worry about such things and, in any event, he was well versed in keeping his own counsel should situations arise where the wrong word could have unfortunate results. He continued down the path of innovation and in 1846

obtained a patent for vulcanizing rubber in or upon moulds, plates, or forms, and retaining such articles under pressure or upon such forms during their vulcanization, by which means permanency of form is given to articles such as shoes, surgical bottles, valves, &c. Impressions from engraved plates so produced remain permanently raised. I may here mention that in this specification I described the hard or highly vulcanized rubber as one of the materials of which I formed my moulds. Although the means of producing this hard material was described in my first vulcanizing patent of 1843, this was, I think, the first time the actual application of it to any particular purpose had been published.

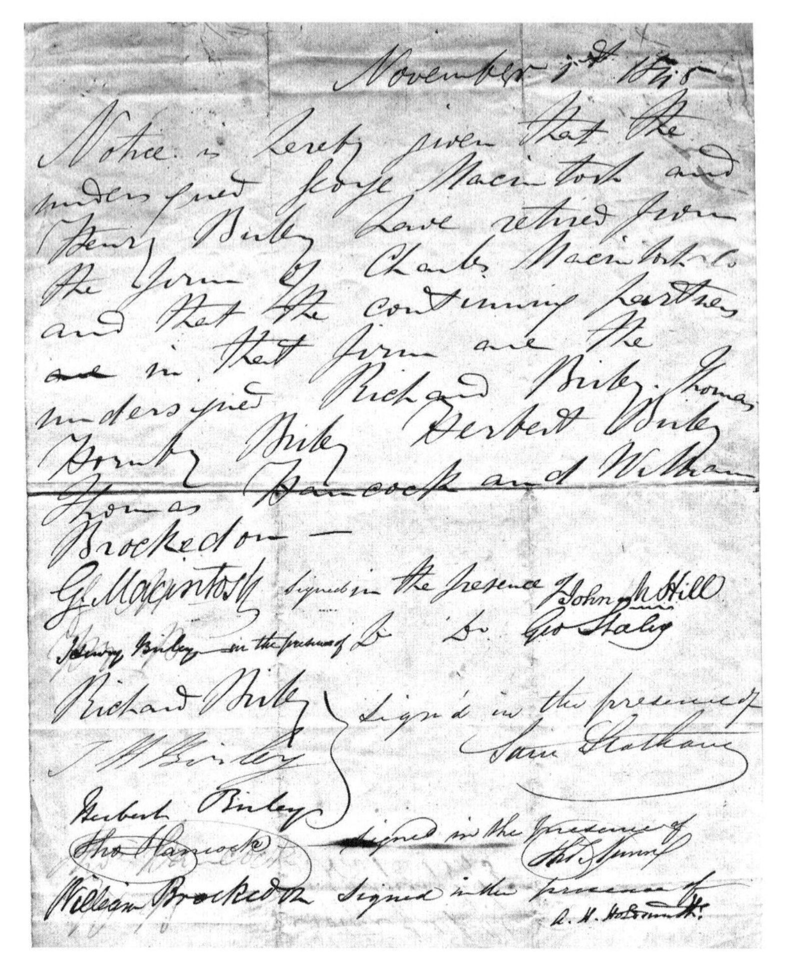

November 1st 1845

Notice is hereby given that the undersigned George Macintosh and Henry Birley have retired from the Firm of Charles Macintosh & Co. and that the continuing partners in that firm are the undersigned Richard Birley, Thomas Hornby Birley, Herbert Birley, Thomas Hancock and William Brockedon —

G. Macintosh signed in the presence of John M Hill

Henry Birley in the presence of Do Do Geo Staley

Richard Birley
T H Birley
Herbert Birley } Sign'd in the presence of Sam Statham

Thos Hancock — signed in the presence of Thos Jenny

William Brockedon Signed in the presence of R. H. Holdsworth

Fig. 7.3 Notice of dissolution of partnership.

An almost endless variety of productions has sprung from the applications of this patent, as the most delicate chasings, tracings, and other ornamental decorations, as well as medallions, bas-reliefs, type for letter-press printing, surgical uses, toy-balls, both solid and hollow, moulding for a great number of mechanical purposes, some of which I have mentioned before, and others are almost daily occurring, some of them requiring a good deal of skill in the formation of the moulds, and experience in adapting and moulding the materials for the requirements of inventors and patentees, particularly in articles which

Fig. 7.4 Ebonite pipes showing the detail obtainable on their surfaces.

were to be hollow, and yet to present on the external surfaces letters, figures, and embossed patterns.

It is particularly interesting to note that he used hard rubber or ebonite to make his moulds. An initial master could be made from wood or metal, depending on the detail and surface finish required, using established technology and the ebonite mould taken from this (Figs 7.4 and 7.5). It would take the finest detail of the master and transfer it to the soft rubber article.

Indeed the range of ebonite articles on the market before the advent of cheap plastics was almost as great as that of soft vulcanized rubber. Thomas described some of them:

The hard vulcanized rubber has been applied to many useful purposes to which this patent has contributed. Combs, knife and other handles, ornamental panels for carriages and furniture, stop-cocks, tubing, pump-barrels, pistons and valves for use in chemical works, &c. &c.—these are capable of being turned in the lathe, and to have screws cut on them in the same manner as is practised with wood, ivory, or metal. I have also had some flutes made of it, the colour is a jet lack, and it polishes like ebony; the notes or sounds are equal to the best flutes, whilst they are said to be produced with greater ease by the performer. I furnished the material to the flute maker without instruction, and he made it in his ordinary practice. This patent for moulding and giving permanency to forms by vulcanizing has, in fact, tended to bring the hard 'changed' rubber prominently into view; and an endless variety of articles have been and may be made by means of this patent, besides the uses to which it is being applied as

Fig. 7.5 Detail of the boy on the fish in Fig. 7.4.

the raw material, from which to manufacture various articles.... There is no other limit to the extent of its employment, except that which arises in point of utility, appearance, or economy in competition with other materials. The turner, the engraver, the comb-maker, and most other artists and mechanics, have only to apply their ordinary means, tools, and skill, as to wood, ivory, metal, and other substances.

There are several copies of an ebonite medallion (Fig. 7.6) extant which show a bust of Thomas Honcock and although it cannot be read in the print, the writing to the right of the medallion expresses Thomas' pride in having taken patents out during the reign of three monarchs; George IV, William IV, and Victoria. He only missed it being four by three months to the day with his first patent of 29 April 1820, shortly after the death of George III.

Two major problems remained in the story of vulcanization. One would not be solved until the 1920s,[3] but the other would—and almost immediately! Vulcanization required high temperatures and therefore the chemicals which could be added to the rubber before vulcanization were restricted to those which were stable at those temperatures, in essence naturally coloured mineral fillers. Thomas takes up the story:

In the spring of 1846 a patent was taken out by Mr. Alex. Parkes for a process by which he produced in rubber nearly the same effects or properties possessed by vulcanized rubber, and gave it the name of *'converted rubber'*. His process is an elegant and simple one, and consists in immersing the rubber in a solution of the chloride of sulphur in bisulphurate of carbon, or pure coal naphtha

Fig. 7.6 Ebonite medallion showing Thomas Hancock.

cold, no heat being required; a thin sheet of rubber is by this means '*converted*' in a minute or two, and when dry is found to have acquired the properties of insolubility at ordinary temperatures, and to be insensible to cold. The process is capable also of producing the horny state, similar to hard vulcanizing. The results effected by this process are as mysterious as those by vulcanization; they could not have been anticipated, they cannot be accounted for, and they prove moreover that heat is not absolutely necessary to produce this obscure change in the condition of rubber.

With his usual honesty he then added that '... the mention of this patent would be hardly consistent with a "personal narrative," but for the reason that I assisted Mr. Parkes in his experiments and specification'.

Charles Macintosh & Co. purchased the patent and was then able to incorporate plant dyes (and within a few years the new synthetic chemical dyes) which would not have withstood the high-temperature vulcanization process and so could choose the colour of their products to suit market demands. It was soon discovered that the reaction could be carried out in an atmosphere of sulphur chloride without the need for solvents and all the problems that related to the use and removal of them. The process is still used today and one domestic product where its usefulness can be seen is brightly coloured and decorated ladies' bathing cap.

It is interesting to note Thomas' friendship with Alexander Parkes who was some 30 years younger than he. Unlike Thomas, he did marry and had eight children by his first wife, remarrying after her death to sire a further 12! As well as his 'cold' vulcanization process, he introduced a new material which he called '*Parkesine*'[4] at the 1862 International Exhibition in London for which he was awarded a prize medal. *Parkesine* was based on cellulose nitrate and is considered the first successful semi-synthetic plastic.

It would seem that Thomas was also anticipating the world of psychedelia when he wrote:

> These converted surfaces also print well, and the most delicate impressions from copper-plate engravings are produced upon them. Gutta percha and compounds of this substance with rubber are equally susceptible of improvements in the same way. Observing that the elastic web imported from France was ornamented with designs in various colours effected by expensive and peculiar modes of weaving them, the thought struck me that patterns might be printed on them, and on trial, I found this could easily be done, and some novel results ensued. The web could be printed of any given pattern in its contracted state; and then, on extending it, all the objects in the pattern were extended in due proportions until it became like a dissolving view almost lost to the sight. On allowing it to contract, the lost pattern and colours were restored in all their freshness; if, on the contrary, the printing was done whilst the web was in its extended state, a curious concentration or condensation of the pattern was produced.

Thomas was soon to come up against the man who had brought sulphur and rubber together so successfully: Nathaniel Hayward. He had stayed in the rubber business although, perhaps unkindly, he had had to pay Goodyear royalties for using his American vulcanization patent and, still in Connecticut, he founded the Hayward Rubber Company. The use of rubber in the early part of the nineteenth century provided a clear divide between England and America. In the former, possibly because of its wet climate, there developed a passion for the mackintosh while in the latter it was the rubber overshoe which held pride of place. In 1847 Thomas was advised that some American vulcanized rubber overshoes were being placed on sale in England. They were being manufactured by the Hayward Rubber Company and after some argument whether Thomas' patent was being infringed the American company accepted that it was in the wrong and Thomas granted it an exclusive license for the importation of American overshoes of their own manufacture 'on moderate terms, which license exists at the present'.[5] There is no reference in Thomas' *Narrative* of any bad feeling coming from across the Atlantic and he was to write in pleasant and complimentary terms to Hayward when sending him a copy of the *Narrative*.

8

Gutta Percha Comes to Town

The discovery of gutta percha and its arrival in London—suggestion by Michael Faraday that it could be used to insulate underwater cable—Charles Hancock's interest aroused while working with Walter to produce an artificial bottle cork—involvement with Henry Bewley, soda water manufacturer—machinations all round end with Charles being outmanoeuvred by Bewley and new partners—Walter reduced to a paid employee and then dropped—Charles also kicked out. New company formed by Charles and (reluctantly) Walter—battle joined with furious price cutting resulting in bankruptcy for the new company—Charles fights on to no avail while Walter surrenders to ill health and dies—obituary. The rump of the business sold to S. W. Silver, whose company becomes one of the most successful in laying undersea electric telegraph cables. Sir William Siemens and the German input.

At the same time that Thomas was developing the vulcanization process, a new material arrived on the shores of the United Kingdom. It was brought from the Far East by Dr William Montgomerie and was called by its Malay name—gutta-percha.[1]

It had actually first been introduced to Europe by the Tradescants in the middle of the seventeenth century. The two Tradescants, father and son who were both named John, were among the earliest English botanists and plant collectors. John the Elder began his career as head gardener to the Earl of Salisbury at Hatfield House, Hertfordshire, and journeyed as far as south as north Africa and north to Arctic Russia in search of new plants. He bought a house in London in 1625, when he was gardening for Charles I, which he called The Ark because, as well as living there, he used it to display the collection of curiosities he had gathered from around the world. This attracted so much interest that he opened part of the family home as a museum. John the Elder died in 1638 and his son, John the Younger replaced his father as gardener to King Charles and followed him in his life's work. Neither of the Tradescants had ever visited the Far East, but in a publication dated 1656 John the Younger refers to a 'plyable Mazer wood, which when warmed in hot water will work to any form'. This is considered to refer to gutta percha since no other material suitable for making mazers (goblets) which will soften in hot water was known[2]. It had presumably been gifted to the museum by some unknown collector.

Gutta percha has the same chemical make-up as natural rubber but the way in which the molecular structure is arranged is different, and this affects its properties. Instead of being elastic, it is rather like a block of hard plastic at normal temperatures but when heated in boiling water it softens and can be moulded to any required shape. Its other important property is that its dimensions are virtually unchanged as it cools after moulding—most materials shrink and this has to be allowed for in complicated or crucially shaped components. It is also impervious to water, a crucial property for its main application in the years to come.

Dr Montgomerie saw how the Malay workers made handles for their machetes with it, and it struck him that there was some potential for its use both as knife handles and for various medical devices. After some experimentation he prepared a report for the Medical Board of Calcutta and copied the documents to The Royal Society of Arts in London. This latter organization was not what its title implied. It had been founded in 1754 by William Shipley, Viscount Folkestone, and Lord Romney, and its main objectives were to promote the arts, productivity, and trade. It was the first organization in Britain with these broad terms of reference, and it held weekly meetings which were attended by a cross-section of respected individuals from all sections of London society and business. Needless to say, Thomas, Walter, and Charles were all members, as was their friend Michael Faraday. In 1844 the Royal Society of Arts (Joint Committee of Chemistry, Colonies, and Trade) resolved that 'this substance appears to be a very valuable article and might be employed with great advantage in many of the arts and manufactures of the country',[3] and Faraday suggested that the impermeability of this new material to moisture could make it ideal for insulating the electric cables[4] which were spreading round the country as the electric telegraph plodded inexorably on and could prove invaluable when the desire for it to cross the English Channel arose. At this point Charles Hancock (Fig. 8.1) got to hear about it.

Fig. 8.1 A very early photograph of Charles Hancock taken in the early to mid-1840s.

Charles was now among the finest and most patronized animal painters in Britain. Finding that portraits were not the source of wealth he had once hoped for, he had discovered a more lucrative market. This was the age in which many a man would be happy to pay a hundred guineas for the painting of his favourite horse when he was reluctant pay £10 for a portrait of his wife, so Charles had specialized in horses. In 1830 he had taken a studio at Tattersalls, the bloodstock dealers.[5] Where better to find patrons who had just acquired the object of their desires? This resulted in a string of commissions, which fully occupied his time and took him to the top of his profession. Still he was not satisfied and he continued to seek some industrial venture. He had thought that he had found it in 1838 when he applied for a patent for printing lithographic drawings in relief, but he found no time to develop it. Thus, when Hullmandel applied to patent a similar process in 1840, Charles, who was as good at spending money as fast as, if not faster than, he could earn it, was reduced to issuing mere impotent words through the columns of the Art Union to try to prevent the perceived infringement, adding bitterly:[6]

On seeing the specification of Mr Hullmandel's patent I immediately gave him notice of the infringement and subsequently applied to the court of chancery for an injunction to restrain him. In the commencement of the action I was assisted by the kindness of a friend, who withdrew his support because he did not think the ultimate advantages to be derived would be commensurate with expenditure necessary to maintain and establish my right. On learning this, my first impulse was to sacrifice all I possessed to defend my own property, but in justice to my numerous family, I have stayed proceedings; but it is my firm determination to proceed with this action so soon as my pecuniary means enable me to do so, which period I have good reason to hope is not far distant, when I have no doubt a jury of my countrymen will establish my just rights.

Did he learn from his experiences of taking action through the courts? Later evidence suggests not.

By a remarkable coincidence, at the time that gutta percha arrived in England Charles Hancock and William Brockedon were working to invent an artificial bottle cork, and Walter, as ever, had made a 'very beautiful piece of machinery'[7] which could finely shred rubber and other substances such as cork. A patent was taken out in Charles' name for the process and in Walter's for the machine, and, either learning from, or after advice from, Thomas, the patent included every possible additional material they could think of—including gutta percha. When a test cork was made using this it worked perfectly and came to the notice of Henry Bewley, a soda-water manufacturer from Dublin, who was frustrated by his inability to keep his ebullient liquid in its bottle. Charles sensed the opportunity to make some real money at last, and in February 1845 he hastily signed an agreement with Bewley to work the patent jointly.[8] Charles' idea was to get the cork-manufacturing operation off the ground and then expand into the many other areas where he could see a future for the material. Walter would be fully occupied in designing and making the new equipment, something he was more than happy to do, and of course there would be no problems with this arrangement as the two had worked together for many years to the satisfaction of them both. The new company was called The Gutta Percha Company.

Unfortunately they had a competitor, another member of the Society of Arts called Christopher Nickels. He was well known to the Hancock family, as he had worked with Thomas in Goswell Mews, and was one of the people who went to France to install the factory for Messers Rattier and Guibal in 1828. He now had his own rubber factory in Lambeth with a partner, Charles Keene. The prospect of a rush for patents and the inevitable litigation which would follow was seen by all parties to be only to the advantage of the legal fraternity, and it was quickly agreed that patents would be taken out jointly and expenses shared, although this would not apply to the patent which Charles already held jointly with Bewley.

In his haste to receive hard cash from Bewley and to ensure the agreement was signed without delay, Charles allowed Bewley to insert many clauses into this and subsequent agreements that he was to repent at leisure. There is no other interpretation that can be put on what followed but a tale of lust for wealth, perhaps on both sides, but the gutta percha saga does not reflect well upon Charles. It started with an act of great meanness towards Thomas, for which Charles was never to be forgiven. In a letter to Bewley in September 1845 he reports that he just returned from laying down the foundations of what he considered to be his master patent for gutta percha: a patent that would prevent any one else being able to use the substance without first having to obtain a licence from himself and Bewley. He continued gleefully:[9]

My brother T has no patent for G.P.—he cannot possibly get by our opposition—nor do we hear of any other party...I do not know how much my brother gave for his lot—I saw it at their manufactory the other day—it is of beautiful quality—principally in large round cakes like cheese. He cannot use it without he takes a licence from us—this of course he is not at present aware of—he expects to do great things with it, and has had from the best authority assurance that any quantity may be obtained.

In his reply, Henry Bewley, who might have been wise to pause and reflect upon the character of his new partner who so ill-treated his own brother, wrote that he had come across an article written by Thomas on the subject in the *Gardeners Chronicle*. This had raised doubts in his mind.

About two years ago some account was given of a new vegetable substance called gutta percha. This new material gives promise of considerable importance, as connected with various manufactures...a substance containing such qualities and capable of being wrought with such facilities, will be brought into use for a thousand purposes. The country is indebted to Dr Montgomerie...for this boon which is the more valuable, as he gives it, together with such information as he has gathered, **gratuitously to the public** without any reservation for his own benefit....

Charles, hastily seeking to quell these inconvenient fears, replied[10]

...I regret very much on my brother's account that he has made the publication in the Gardeners Chronicle, because it is obviously made in a bad spirit, and just after he had been beaten at the opposition—had he known of the existence of the mastics and cement patent, he would never have done it, because he would know that it would avail him nothing. He is quite right in saying that Dr Montgomerie made the nation a free present of gutta percha. But in his statement to the Society of Arts two years ago, he did not tell the nation how to clean it—masticate it—make it into thread—artificial thread—waterproof cloth with dissolving it—to make it into boots and shoes—how to bind books with it—to saturate cordage, and various other applications of it. Therefore, for a mode or modes of manufacturing the substance, and also the application of it to various purposes, he left freely open to any one to secure by patent, and one of the very first who took advantage of it...was Thomas Hancock. No Sir, we have no need to fear any publication that has appeared. There has been *'no public and general use'* made of the substance, it has only been used experimentally, therefore all patents are valid patents, and I have no doubt that Charles Macintosh & Co will be large customers to our patents, pertticularly [sic] when they find they have been taken out by practical people.

Now we can see much more clearly the treacherous game he is playing. Not only did Charles hasten to lay down his master patent to try to prevent Thomas from getting any benefit from gutta percha, but he also had his eyes fixed on the lucrative business he saw arising from dealing directly with Charles Macintosh & Co., thus circumventing Thomas. This might even raise the possibility in the corporate mind of Charles Macintosh & Co. that Thomas could be in collusion with Charles, working against the company's interests. In this way two rifts in otherwise healthy relationships might be created, the one between Charles and Thomas was unavoidable, while the potential for one between Thomas and Charles Macintosh & Co. would be an extra bonus. Bewley was eventually disabused of his fears, and an agreement was signed enabling Hancock and Bewley to proceed to the manufacture of products from gutta percha.[11]

A manufacturing base was needed and a large shed was quickly found in Stratford which seemed to be adequate for their initial experiments, and Charles managed to get Bewley's agreement to pay Walter the princely sum of £3–10s per week for looking after the machinery. Their early products—driving belts for machinery, and shoe and boot soles—convinced Bewley that it was time to expand into larger premises. In order to do this he suggested to Charles that a new partner should be found to provide the necessary capital—a suggestion eagerly endorsed by Charles, who, in view of the larger amount of time he would have to spend at the factory, felt in a position to demand some recompense for the time he would be away from his painting. It was agreed by all the parties that he would receive £800 per annum as well as his share of royalties from any patents.

Prospects looked good, while Thomas, who would in November 1846 take out a patent in conjunction with William Brockedon, for 'the manufacture both of rubber and gutta percha, either separately or combined, in all proportions', kept his distance!

Thomas had made a wise decision, as the two brothers seemed to have learned little over the last 20 years. Walter was simply an innocent abroad, happy building his mechanical masterpieces, doing the research and preparing the drawings for proposed patents, while Charles still let his desire for money override any caution that he might have felt about his business abilities. The time was ripe for a repeat of the rubber/tar saga. Business was booming and it was obvious that the existing premises were too small to cope with the demand. The new partner was a Mr Reynolds, introduced by Samuel Gurney, who was senior partner in one of the leading banking houses in Europe. Instantly disliked by Charles, he nevertheless had vast amounts of money and the strength of personality to go with it. The factory in Stratford was abandoned and extensive new properties were leased at 18 Wharf Road, City Road. Reynolds' financial input enabled him to insist that the demand for new equipment be taken out of Walter's hands and placed with larger organizations and, as Walter was only a contractor, he could easily be dispensed with. Walter's role was reduced to looking after the machinery for £250 per annum. Charles admitted his lack of business acumen to Bewley when he wrote defensively:[12]

I consider, to Walter's prejudice and my own, they may have cast our powers in the shade, being ignorant in the practice of the business in proportion as we are unskilled in the ways of the world.

Bewley was a pragmatist and, having all the equipment he needed, as well as part ownership of their joint patent, he was not going to let sentiment interfere with his desire to make money. The clouds were gathering.

The position was to get worse. Bewley's interest in gutta percha had led to his developing a machine to extrude gutta percha tubing and rods which had then been modified by the Hancocks, while working for the company, to coat electric telegraph cables. The patent had been taken out in Charles' name but the discovery of this vitally important process had been concealed from Bewley and the Company. In 1848 the company received its first order for an underwater telegraph cable and from that point demand rocketed to such an extent that it quickly became the company's greatest source of income. There soon developed a furious argument between the Charles and Bewley over who had the right to profit from the Hancocks' modification. Charles contested Bewley's right to use the equipment at all and wrote in anger:[13]

You once told me that I wanted to be a rich man, if this was really the case I should have left the service of the company before now.

When he tried to prevent Bewley using the machine, the latter retaliated by sacking Charles and Walter with one year's salary as compensation. Samuel Gurney, now himself a partner in this potentially prosperous business, raised no objection.

Charles and Walter retired to lick their wounds. Walter was in despair as this would leave him penniless once again, but Charles was prepared to fight on with the gutta percha patents he still owned. He persuaded Walter to join him in a new venture and so they set out once more with their own new company based in Stratford and close to Walter's Stratford works which they called 'The West Ham Gutta Percha Company'.[14] To ensure they had someone working with them who had useful and practical knowledge of the various manufacturing processes, they secured the services of Frederick Farmer Moore, who had been working for them in their previous company. He and Walter were to be partners in the new business with Charles providing the capital but remaining an invisible partner, although he reserved the rights of three of his sons to become partners when they came of age. Charles' sons, Walter, Robert, and John were keen to work for the new company, which became affectionately known throughout the neighbourhood as the Walter Hancock Company. Throughout this turbulent period Walter was subject to increasing bouts of illness, being frequently absent and confined to his bed, while the news that his older brother William had dropped dead in Gravesend while seeing off more of his children to Australia did nothing to cheer him.[15]

The relationship between the two brothers and Charles Bewley was obviously not a cordial one, and open warfare between them was declared with a price-cutting battle. In 1849 the Gutta Percha Company fired the first shots by advertising, under the banner heading shown in Fig. 8.2, the wide range of products it manufactured, with the superb quality and importance of almost every item being vouched for by an independent testimonial.

One of the most unusual sections concerned gutta percha chamber utensils (Fig. 8.3) and speaks for itself—although the last section would benefit from an illustration!

Many advertisements of the time benefited from the absence of the Advertising Standards Authority but the company must be given credit for honesty in this instance as Mr Lawrence's letter still exists and is quoted verbatim (Fig. 8.4).

This obviously caused not only consternation but also anger in the camp of Charles and Walter. In an advertisement[16] of 1 June 1850 we find:

18, WHARF ROAD, CITY ROAD, LONDON.
WHOLESALE WAREHOUSE FOR THE NORTHERN COUNTIES, 2, YORK STREET, MANCHESTER.

The GUTTA PERCHA COMPANY beg to invite the attention of the public to their various Manufactures.

Fig. 8.2 The Gutta Percha Company advertising header of 1849. East Sussex Archives ref. DH/B 136/413

GUTTA PERCHA CHAMBER UTENSILS.

These chamber utensils are strongly recommended for Prisons, Workhouses, Lunatic Asylums, Hospitals, Shipboard, &c. The following testimonial will shew how they have been approved in the Dorchester Castle.

(*Copy*)

DORCHESTER CASTLE,
May 26th, 1849.

I most willingly record my conviction that no article that has been hitherto made use of as a chamber utensil for the use of a Prison so completely answers the purpose as the Gutta Percha.

Independently of being quite *free from any disagreeable smell*, and unaffected by acids, the Gutta Percha utensils are so perfectly light that they cannot be made use of as weapons of assault, and however violently they may be dashed upon the floor cannot be broken. These utensils are now partially in use in this Goal, and it is my intention to request the permission of the Visiting Magistrates to obtain a further supply.

J. V. LAWRENCE, GOVERNOR.

Lancaster Castle, *22nd June*, 1849.

Gentlemen,

I have much pleasure in stating that the experiment made in this Prison of your Gutta Percha Chamber Utensils is very satisfactory; and I hope that the Magistrates will, after some further trial, deem it expedient to bring them into general use throughout the establishment.

I perfectly concur in the opinion of the Governor of Dorchester Castle, that as an offensive weapon, the Gutta Percha article is much less formidable than a wooden or earthen one, and is, on *that account an important desideratum in a Gaol*.

I am, Gentlemen, your very obedient servant,
JAMES HANSBROW, Governor.

It is important to know that these chamber bowls can be quickly converted into life buoys in the event of a shipwreck, and are therefore valuable on shipboard.

Fig. 8.3 Gutta Percha chamber utensils. East Sussex Archives ref. DH/B 136/413.

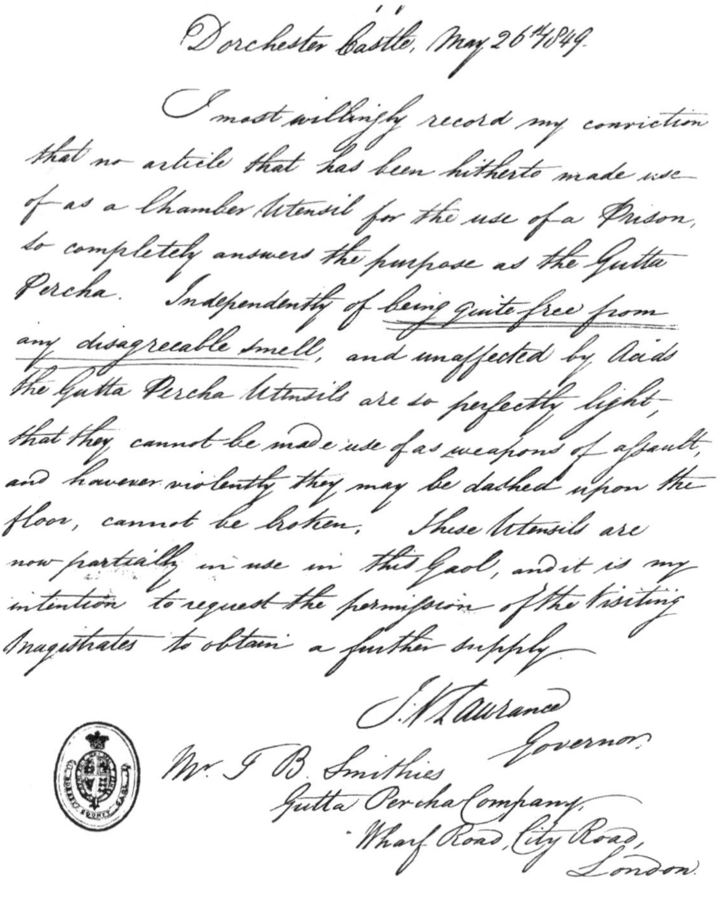

Dorchester Castle, May 26th 1849.

I most willingly record my conviction that no article that has been hitherto made use of as a Chamber Utensil for the use of a Prison, so completely answers the purpose as the Gutta Percha. Independently of being quite free from any disagreeable smell, and unaffected by Acids the Gutta Percha Utensils are so perfectly light, that they cannot be made use of as weapons of assault, and however violently they may be dashed upon the floor, cannot be broken. These Utensils are now partially in use in this Gaol, and it is my intention to request the permission of the Visiting Magistrates to obtain a further supply.

J. N. Lawrance
Governor.

Mr F. B. Smithies
Gutta Percha Company,
Wharf Road, City Road,
London.

Fig. 8.4 Letter to the Gutta Percha Company from Mr Lawrence. Dated 26 May 1849. East Sussex Archives ref. DH/B 136/412

The West Ham Gutta-Percha Company; Charles Hancock's Patents;
Gntlm.
We beg to inform you that the business heretofore carried on by the firm of Messrs Hancock & Co. [this must refer to The Gutta Percha Company but it is not made clear!] has been taken by us, and we respectfully solicit the favour of your support. As we shall have the exclusive benefit of the personal assistance of Mr Charles Hancock, the original introducer of Gutta Percha into use in the manufacturers of this country, and the Inventor and Patentee of all the most important processes connected with it, we shall be able to execute all orders with which we may be favored with a degree of perfection not to be

hoped for from other persons, possessing only an imperfect knowledge of the properties of this valuable material.

All articles supplied from this manufactory will not only be of the best quality of material, (on the purity of which much depending) but of the best workmanship and finish.

Dealers will be treated with on a liberal and uniform system—Such as shall conduce equally to the protection of the trading capitalist, and the encouragement of enterprise.

Every description of Gutta Percha Goods in general demand are kept ready for delivery, including Shoe Soles, Driving Bands, Sheet etc etc.

Our superior and highly approved process of covering Telegraphic Wire, and of moulding tubing of unlimited lengths, without a joint, enables us to execute orders for such articles to any extent and with great dispatch.

Orders for moulded articles of any form and colour, executed at moderate prices, and with superior sharpness and beauty.

Hancock's Patent Bosses which have been so extensively used, continue to be preferred by the Flax Spinners of England, Scotland, Ireland, France, Belgium and throughout the continent—and we feel assured from the recent improvements made in them, the superiority of their manufacture, as well as the expeditious execution of orders, that the demand for them will continue to increase.

However, given the limited resources of the Hancocks' company, the more it managed to sell, at 25–35% discounts, the more money it lost and it was inevitable that they would eventually lose the war as well as individual battles. The company went into bankruptcy and Charles took over the bankrupt stock, spending further good money on pursuing those through the courts who still had outstanding debts to their former company. Walter finally surrendered to despair and ill-health and, in spite of Rebecca's comfort and care, died in 1852.

Friday morning last, at the deceased's residence, West Ham Lane, Mr Hancock had an attack of gout, a complaint to which he was subject. It seemed he retired to rest at his usual time on Thursday evening and on the following morning a letter arrived by the first post, which was taken to him; he perused it in bed where he was subsequently found in a sitting posture, with his eyes closed. Mr Hancock's medical gentleman was called in, but could render no assistance as the vital spark had fled.[17]

Rebecca remained in their house with Elizabeth Ann until her own death in 1865, and later that year Elizabeth married James Bennett. She lived until 1911, when she died in Maidenhead, having bequeathed to the world five children: four girls and a boy.[18]

Walter is remembered today only as the inventor of steam vehicles, although this was considered at the time to be a relatively minor part of his total achievement. This is reflected in his obituary,[19] which also takes a less than subtle side-swipe at those partners he was unfortunate enough to become involved with.

About this period Mr. Hancock invented a steam locomotive for common roads, and probably the non success of this undertaking may be mainly attributable to the formation of the railways, which superseded an invention, which to say the least, displayed considerable skill and ingenuity, but yielding but little pecuniary compensation to the projector. Mr. Hancock during his life had made many important discoveries in that particular branch of science to which he had directed his attention, but the pecuniary benefits arising therefrom were chiefly obtained by others whose capital and business habits afforded them ample opportunities for the accomplishment of that object.

The week after his death his friends inserted a short obituary in the local paper[20] which gave a brief résumé of his life and concluded:

A numerous circle of the deceased's friends are desirous of recording their tribute to the memory of one whose kind and amiable disposition—whose modest and unassuming manner justly entitled him to their respect and esteem, and now his remains are consigned to the cold and silent grave, his memory will long be cherished in their affections.

He was—
But words are wanting to say what;
Say what a man should be, and he was that.

Oddly, no trace of exactly where he was buried in Bow Cemetery has been found. He lies in humble obscurity, a fate undeserved by one of motoring's earliest and most distinguished pioneers.

Charles however was still determined to gain what he saw as his right, and in 1855 we find the remains of company based at 18 West Street, Smithfield (see Fig. 8.2). He was also using the law as this Law Report in *The Times* of 19 May 1855 shows:

The Times, Saturday, May 19, 1855;
GUTTA PERCHA GOLOSHES.—Notice.—The oppositions to the sealing of the PATENTS for MAKING our IMPROVED GOLOSHES, entered at different stages of its progress on behalf of Henry Bewley, of Dublin, and Samuel Gurney, of Lombard-street (trading under the style of the Gutta Percha Company, at the Wharf-road.) and Christopher Nickels, of Goldsmith-street, have been this day over-ruled by the Lord Chancellor's final judgment, with costs.

Notice is hereby given, that the sole authorized manufacturers of our patented golosh are the West Ham Gutta Percha Company, 18, West-street,

Smithfield, London, by whom wholesale orders to any extent are executed; and further, that all persons who make or vend the same, or any imitation thereof, other than those bearing the stamp of ourselves, or the West Ham Gutta Percha Company, will be forthwith proceeded against in such manner as we may be advised for infringement of our patent rights.—Dated May 6, 1855. MARTIN and HYAMS, Union-street, Bishopsgate, Patentees of the Improved Golosh.

The illustrations in Figs 8.7 and 8.8 were taken from a broadsheet publication illustrating many of the products in the catalogue, the front page and index of which are shown in Figs 8.5 and 8.6.

Finally, however, Charles had to admit defeat. His dispute with Bewley was heard in the Court of Chancery in 1855, where his suit was dismissed with costs. Unwisely deciding to take the matter to appeal in 1860 he lost again, also with costs, which, although not specified, must have been considerable. He was left with very little to dispose of but did,

UNDER

CHARLES HANCOCK'S PATENTS.

THE ORIGINAL PATENTEE AND MANUFACTURER OF

GUTTA PERCHA.

LIST OF PRICES

OF ARTICLES MANUFACTURED BY

THE WEST HAM

GUTTA PERCHA COMPANY,

18, WEST STREET, SMITHFIELD,

LONDON.

Nov. 23rd., 1855.

R. LINDLEY, PRINTER, OLD BAILEY, LONDON.

Fig. 8.5 Cover of the 1855 catalogue/price list for The West Ham Gutta-Percha Company.

INDEX.

—oo—

Fig 8.6 Index of the 1855 catalogue/price list for The West Ham Gutta-Percha Company.

Angel Inkstand.

Fig 8.7 'Blow-up' of the inkstand shown on the page of illustrations.

eventually, enter into an agreement with Mr Samuel Winkworth Silver, which resulted in the formation of 'The Indiarubber, Gutta Percha & Telegraph Cable Works', an extension of Silver's rubber company which was located near the Royal Docks. So many of Silver's employees lived

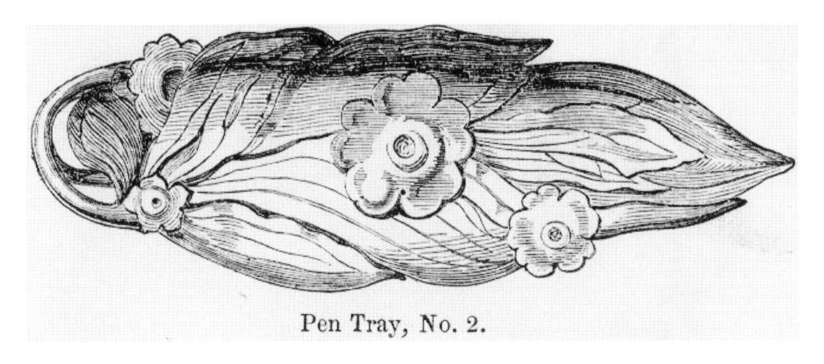

Fig 8.8 'Blow-up' of the pen tray shown on the page of illustrations.

there that it became known as Silvertown, a name which defines the area to this day. Charles had, by now, lost interest in gutta percha, but Silver and the company had his patents and decided to go into the booming electric telegraph cable-laying business with a vengeance. The first cable manufactured and laid by the company was for the Submarine Telegraph Company running from Dover to Cap Gris Nez, which was laid in 1865. This was followed in 1867 by a cable which linked Havana, Cuba—Key West and Key West—Punta Rassa.

The company became active in promoting telegraph companies.[21] It set up the West India and Panama Telegraph Company, the Cuba Submarine Telegraph Company, the Panama and South Pacific Telegraph Company, the Direct Spanish Telegraph Company, and the West Coast of America Telegraph Company. By the mid-1870s the company had four cable-laying ships, the CS *Dacia,* CS *International,* CS *Silvertown* and CS *Buccaneer* but by 1914 only one remained—the CS *Dacia*—which was torpedoed in December 1915. The glory days of gutta percha were over as wireless telegraphy replaced cable telegraphy. One more cable-laying ship, the CS *Silvergray, was commissioned* but it was only used for a year or two before being sold. The focus of the company changed back to rubber and the name was changed to the Silvertown Rubber Company to reflect this. It was later to become part of B.T.R. Industries Ltd, a conglomerate we shall meet again in this story! Charles, who died in 1877, certainly saw the development of his dream and may have felt proud of being, perhaps, the person most responsible for the development of the new industry. Whether he gained anything financially from it is not recorded.

Fig. 8.9 The remains of the Gutta Percha Company Works, City Road.

Bewley's company also concentrated on gutta percha-coated underwater telegraph cables but had a shorter existence, although being in a position to begin operations some years earlier. It supplied them for both the 1850 and 1851 cross-Channel telegraphs, but in 1853 their premises in City Road were gutted by fire. It must have been a spectacular affair as it was reported in the Illustrated London News of 11 June with a sketch of the devastation (Fig. 8.9). The manufactory must have been an impressive complex before its destruction.

GREAT FIRE AT THE GUTTA PERCHA COMPANY'S WORKS

A FIRE of a most destructive character broke out on Sunday morning, shortly before eleven o'clock, on the premises of the Gutta Perch Company, in the Wharf-road, City-road. The Company's premises included an immense warehouse for the raw material; a tube and wire manufactory, in which the wire for the submarine and subterranean telegraphs was prepared; the warehouses for manufactured goods, with a range of counting-houses over them; three boiler and engine-houses; the cleansing and kneading-rooms; the rolling mills; and the mechanics' shops, where the whole of the valuable machinery employed on the works has been made. Of this immense range there now remains only the raw material warehouse, the tube manufactory, and the Mechanics' shop.

The fire was discovered by the wife of a confidential servant of the company, who observed smoke issuing from the warehouse above the boiler-rooms.

The premises contained property valued at £100,000; and it is supposed that the damage done will exceed £30,000. The fire burnt with extraordinary rapidity. On the premises of Mr. Gorton, for making Edwards' fire-lights, a ship-load and a half of wood was destroyed; and so great was the body of fire, that several buildings on the opposite side of Wenlock-basin, 100 feet across, were ignited. Two vessels lying in the canal, near the factory, were totally destroyed. Fortunately, the telegraphic wires and tubes were saved; and three immense tanks of naphtha were saved by the bravery of the fire-brigade. The counting-house books and papers, which were contained in one of Milner's patent fire-proof safes, were also saved, although the intense heat in many parts melted the outside of the case. The property was insured for the most part in the Yorkshire, Imperial, Scottish Union, and Royal Insurance offices.

During Monday a most searching inquiry was made by the brigade authorities, and also by the officials of the company, for the purpose, if possible of ascertaining how the terrible disaster originated, but nothing that could be depended upon could be learned. The only thing definite that could be gleaned was that, when the factory was closed on Saturday, there was not the least smell of fire in any part of the premises, and every compartment seemed perfectly safe as usual.

Such an unusual circumstance as a fire communicating with buildings so far from the one in which it began, on the other side of the Wenlock-basin, appeared to everyone who did not witness it next to an impossibility, more especially so, as there was a large sheet of water to separate one row of premises from those where the flames commenced. Those persons who witnessed the commencement of the conflagration, however, do not express the least astonishment that the flames should have travelled to such an extraordinary distance, for they describe the fire to have shot from the windows and doors like several furnaces in full play. The land side of the factory has been measured, and it was found to be precisely of the same width as the basin—namely, 110 feet. But that space gives a very faint idea of the extent of the whole premises, as scarcely an inch of ground was left unoccupied, the whole being turned to some advantage, either for warehouses or workshops.

Upon a closer examination the whole of the valuable stock-in-trade, &c., in the machine-rooms, the band-houses, the cutting-rooms, the rolling-houses, and the press-houses, and the picture-frame department, appear for the most part destroyed; but there seems some chance that many of the bevel cog-wheels and some other heavy portions of the machinery may with care be worked up again.

Out of an immense number of drinking-cups, plates, fancy frames, inkstands, &c., which were stored in the premises, nothing can now be seen but heaps of partly-consumed gutta percha, which cannot be again worked up.

The company, it is stated, have been accommodated with other premises in the neighbourhood, so that only a temporary interruption of their business will be occasioned. Mr. Statham, the manager, states that he hopes to be able to resume the deliveries of telegraphic wire within a month.

It is not known whether the hopes of Mr Statham were fulfilled or not but the company did not continue independently much longer, and in 1864 it merged with Glass, Elliot, and Company to form the Telegraph Construction and Maintenance Company Ltd.[22]

Another person who was also a friend of Michael Faraday had attended that meeting of the Royal Society of Arts and had appreciated the properties of gutta percha. This was Mr (later Sir) William Siemens, who managed to obtain a sample, and while Charles was busy anticipating, then losing any chance of obtaining, a fortune he also picked up on Faraday's idea and sent some to his brother Werner Siemens in Berlin with the suggestion that he try it as a telegraph wire insulant.[23] It was a great success and Werner's business thrived in Germany. In 1846, he became a member of a commission set up to introduce electric telegraphs in place of the optical ones then in use. He succeeded in

Fig. 8.10 Sir William J. Hooker and Werner Siemens: commemorative plaque made of gutta percha.

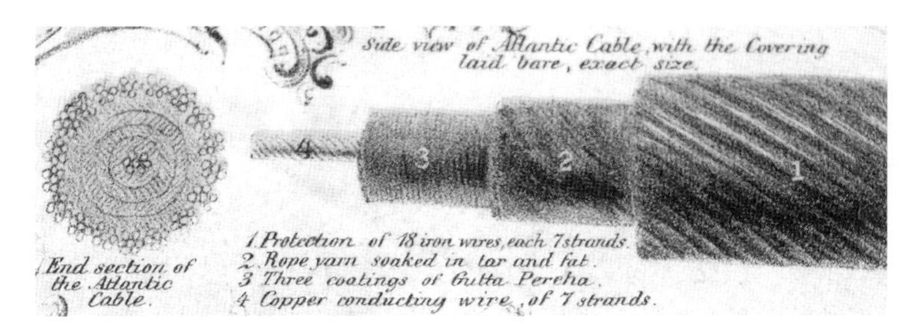

Fig. 8.11 Cross-section and cut-away illustrations of an electric telegraph cable.

getting the commission to adopt underground telegraph lines, and in 1847 he laid the first underground line for the Prussian army. This was followed in 1848 with his first government contract, which required the setting up of a 600-km government telegraph line from Berlin to the National Assembly of Frankfurt am Main. From that point the electric telegraph spread throughout Germany and on through Europe. It is interesting to note that during the 1850s, when the Siemens company moved into submarine cables, it bought its gutta percha-coated cable from The Gutta Percha Company and then used its own factory to armour it before use.

In 1847 the gutta percha plant had been first named and illustrated by Sir W. J. Hooker, then Director of Kew Gardens and the plaque (Fig. 8.10), made of gutta percha, celebrates their relationship. The dates in the bottom centre are difficult to read in reproduction but show '1847–1897'. Both men were dead by 1897 and the dates probably indicate the founding of Siemens' company (1847) with J. G. Halske and either (or both) the half-centenary of the company and the date at which it went public as Siemens and Halske A. G. (1897).

Figure 8.11 illustrates the construction of a gutta percha-insulated electric telegraph cable and is taken from the cover of a piece of sheet music, 'The Ocean Telegraph March' by Francis H. Brown, published in1858, to celebrate the first telegraph cable to be successfully laid across the Atlantic Ocean in that year. It joined Foilhommerum, Valentia Island in western Ireland, with Heart's Content in eastern Newfoundland and was laid by the Atlantic Telegraph Company led by Cyrus Field.

9

The Great Hose Controversy[1]

> A sequence of letters between Thomas and Charles Macintosh & Co., The latter disputes the right of James Lyne Hancock to make certain hoses and Thomas takes up cudgels on his behalf—a rare document written outside his *Narrative* by Thomas setting out in detail the acquisition of his late brother John's hose business by James from Macintosh & Co.—Macintosh & Co. unmoved and insist that no right of manufacture exists—Thomas replies with a vast number of references to the interchangeability of the words 'hose', 'tube' and 'pipe'—no more is heard from Macintosh & Co. The whole argument probably rested on one sentence in the deed of trust between James and Macintosh & Co., which was not mentioned until much later in the argument.

In Thomas' *Narrative* there are virtually no references to life outside his manufactory because, as he wrote in the introduction, 'It is simply an account of my own progress in the manufacture . . .'. While it is reasonable that he did not write about Walter and Charles' experiences with gutta percha, it does leave some gaps in his own activities which, when filled in, provide us with some additional insight into his character.

In the *Narrative* he jumps from 1847 to the Great Exhibition of 1851, but during 1849/50 the question of when is a hose not a hose arose when James Lyne Hancock sent an unusual piece of tubing (or hose?) to be vulcanized at the works of Charles Macintosh & Co.[1] There was obviously some manufacturing fault, something Thomas does not dispute, only saying that this is not unusual with a new design and it would be replaced, but there is also an indication from Manchester that this piece was not a tube within the agreement drawn up when the business was sold on to James. Who wrote the initial letter from Charles Macintosh & Co. or, indeed, the subsequent ones is not known but it was obviously someone who did not know Thomas very well or the history of his relationship with the company. The one person we know it was *not*

Fig. 9.1 James Lyne Hancock's mobile advertisement.

was the wonderfully named Alonzo Buonaparte Woodcock, brother of Edward Woodcock, who was involved in the setting up of the French factory for Messers Rattier and Guibal in 1828. Alonzo had worked with Thomas for many years and at this time was works manager at the Macintosh factory in Manchester.

In virtually all of his writings, Thomas had hitherto appeared calm and in full control of any emotions he may have felt, but the letter from Charles Macintosh & Co. crossed a boundary and, as the argument built, anger dripped from his pen. He obviously resented the somewhat imperious stance the writer had taken but was even more offended by the implied insult to both his and his nephew's character. It should be remembered that during this altercation Thomas Hancock was, and remained until his retirement in 1858, the senior partner in Charles Macintosh & Co. and although he continued to reside in Marlborough Cottage, only a very few miles from his old factory, he travelled regularly to the Macintosh works in Manchester.

He felt the perceived insult demanded his personal attention and responded on 9 October 1849. His reply was a little prickly, initially regretting that several days had elapsed between his first sight of the letter from Macintosh and Co. and his opportunity to talk with his nephew, James, since the matter should have been easily resolvable. He went on to say that there seemed to be confusion and errors on both sides but that:

The letters from Manchester on this subject contain expressions and allegations that should never be admitted among parties having due respect for upright and honourable dealing and should it be charged upon any man of irreproachable character that in conducting his business he is in the habit of deliberately shirking his agreements.... Matters in dispute and capable of explanation should be distinguished from those that arise from dereliction of principle....

He made a valid point that the product had all the appearances of a tube, as the word was generally understood, and it was not up to his nephew James to determine the end use for which it was designed and, indeed, if he had pursued the matter, the client would have been unlikely to tell him, such was the suspicion and secrecy which surrounded any novel applications of rubber at that time! Thomas continued by saying that he was making no defence of his nephew if he could be shown to be in the wrong, but that, in his opinion, there is not a man with more upright intentions nor more industriously and indefatigably attentive to business than he.

This obviously elicited some outbursts from Charles Macintosh & Co., which appear to have been lost—or perhaps, since it was late in the year, thrown on the fire in disgust. Thomas obviously felt that the new junior management did not fully understand the relationship between himself and James with the Macintosh factory, so a few weeks later he sent the company a second letter in which, *inter alia,* references to the missing letter or letters make it clear that peace was not edging its way over the horizon. This letter is reproduced here with only minor editing as it represents Thomas' retrospective view of how the company of James Lyne Hancock came into being. Although Thomas founded his company in 1820 it was James and one other family member who carried it forward until well into the twentieth century.

STATEMENT AS TO INDIA RUBBER HOSE PIPE
Copy sent to C.M. & Co., Manchester
November 1849.

The India Rubber Hose Pipe and Tubing business was commenced more than twenty years ago by my late brother John with my assistance—it was fought by us for years against all sorts of opposition from various interested parties until at length it was firmly established; the principal Breweries and Distilleries gradually appreciating the value of the new Hose and adopting it. My brother on leaving London in 1833 sold the business to Charles Macintosh & Co., (my late much esteemed partners) and such contracts as had been made with the principal consumers were handed to them with the Stock and Tools etc., the workmen also continuing in the employment. This business was then carried on by me in conjunction with my own in Goswell Mews on behalf of the firm until at length the two concerns merged into one together with my premises Stock and Plant and I became associated with the firm of Chas. Macintosh & Co.

In 1842 the trade of Chas. Macintosh & Co. greatly declined and they came to the determination of abandoning their establishments in London as no longer profitable and Goswell Mews with the rest. My nephew J.L. Hancock who had been for some years my principal assistant thought the business might suit him if he could, by his exertions, increase it—he accordingly arranged with Chas. Macintosh & Co. to take the Stock and Plant at a fair value with the lease of the premises and the un-liquidated debt due to me on the Plant—he engaged also to do some few things for Chas. Macintosh & Co. during the continuance of their keeping open their shops in London etc. at a fair and reasonable charge C.M. & Co., agreeing on their part to give up to him the business in Hose Pipes and Deckle Straps with which in future they were not to interfere. In pursuance of this arrangement he paid them in cash between eleven and twelve hundred pounds and they handed over to him the Lease Stock and Plant and contracts with Brewers etc. As had been done by my brother on the former occasion—all orders taken at the shop were sent direct to him and all parties employed by Chas. Macintosh & Co., did the same at Manchester—a circular was also addressed to customers in the usual way announcing the change of proprietorship in the business—undisturbed possession of the concern was in this way secured to my nephew (as far as Chas. Macintosh & Co. could do it).

In 1843 I took out my patent for vulcanizing which soon revived the business of the firm and ultimately raised it to a prosperous and flourishing condition. When I arranged with my partners respecting this Patent I threw all the advantages of it freely into the concern only reserving to myself the sole use of the patent as far as regarded the manufacture of Hose Pipes, Billiard Cushions and Deekle Straps which I did chiefly with the view of securing to my nephew the business he had purchased and to which he had now devoted all his energies as the business of his future life. Since that time he has been at great expense both of money and labour to increase his trade in which he has to a certain extent succeeded. I have given my nephew a sole license to make the reserved articles. From this statement four principal points are clearly established.

First—that the India Rubber Hose business originated not with C.M. & Co. but with my brother John and myself.

Secondly—That my brother John sold this business to C.M. & Co.

Thirdly—That as far as regards Hose Pipe, Deekle Straps and Billiard Cushions I possess a property in Vulcanizing Patent as fully as if the Patent were still wholly in my possession.

Fourthly—That Chas. Macintosh & Co. sold or in some way or other transferred the Hose business to J.L. Hancock on condition of his taking all its liabilities Stock Plant etc. etc., for which he paid them in hard cash from £1100 to £1200.

The three first mentioned points are incontrovertible. The fourth has within the last few days past been questioned by C.M. & Co. How justly or otherwise is not our present enquiry. I take it for granted then that the four points thus stated are indisputable and at all events the Hose business was in some way or other transferred by C.M. & Co. to J.L. Hancock and that for nearly seven years this manufacture had been his peculiar business and in which as far as Vulcanized Hose is concerned he has had no competitor. Now J.L.H. standing thus, it is clear that he would be wanting to himself in an extraordinary degree if he did not take every prudent measure for the maintenance of his position, particularly under the circumstances I am now narrating.

I have forgotten earlier to mention that J.L.H. has not any means of manufacturing his own materials nor of vulcanizing his own articles. The first he has always taken from Chas. Macintosh & Co. and hitherto in compliance with their own wishes he has at very great inconvenience and loss to himself sent his goods to Manchester to be vulcanized. For some time past Chas. Macintosh & Co. have made complaints of an incidental nature such as commonly occur in business to J.L.H. most or all of which have either been founded in error or satisfactorily explained—these instead of being treated as such matters generally are—explained and forgotten—have been designated *'breaches of agreement'*, *'contrary to agreement'* and *'we adhere strictly to our agreements'* etc. etc. If instead of thus prematurely arriving at conclusions and adopting expressions neither justified by facts nor sanctioned by usage in the conduct of commercial correspondence the more ordinary course had been pursued matters might have still gone on undisturbed.

Another circumstance connected with this part of the narrative should here be noticed for the convenience of both parties in conducting their business. A mutual concession had been made and long acted upon that in certain cases either party would allow the other to avail himself of their respective privileges under the Patent on condition of paying a certain royalty agreed on—thus a royalty account sprang up—a somewhat doubtful case arose—the royalty on which was under a shilling—both parties contended they were in the right—perhaps nobody could decide which—Chas. Macintosh & Co. however cut the matter short by giving notice that they would allow this accommodation on their part no longer—no retaliation was offered but the inconvenience is still felt—no explanation or reason was given for the proceeding. In the midst of this untoward kind of intercourse a letter was addressed to me on the 9th of October relative to these affairs and containing this portentous passage. *'Perhaps a few months will throw the trade open and thus settle the whole matter'*. I could not shut my eyes to the latent intention construed in these words—but

I might be mistaken. I therefore wrote thus in reply. *'With regard to the Hose business however it may become open to others it cannot be so to us we sold it to my nephew.'* I was told in reply that my view of the case could not be concurred in, and on pressing for an explanation I was told (2 November) that the Hose Pipe business never was sold to my nephew. This was explicit but it was astounding and fully opened up the whole case and showed that the confidence in which we had been securely slumbering was unsafe—it did more—it roused us out of our slumbers. Is this ambiguous? Then let me clear it up—it is alleged that *'Perhaps a few months will throw the trade open and thus settle the whole matter.'* How will matters stand In that case?

Thomas was in no doubt of the implications to his nephew if Macintosh and Co. was able to repudiate the transfer of the hose business to him. Other rubber companies standing in the wings would move in to compete for business while James would find himself with his hands tied and with neither a source of masticated and mixed rubber nor facilities for vulcanizing his tubes. He would be solely dependent on the good offices of Macintosh and Co., who would then be one of his competitors, and they would have it in their power to cut off his supply of compounded material or to refuse to vulcanize it at any moment they chose.

This would effectively bring an end to the company. So Thomas threatened to provide the means to manufacture and vulcanize products in Goswell Mews, only for Macintosh to reply on 7 November that this would be looked upon 'as a declaration of war which will be rigorously carried on our part and every one of our principal men will give their utmost aid to carry it out!!!'

Thomas was still prepared to fight his corner but thought that some conciliatory comments might reduce the temperature and bring the matter to a close. He therefore suggested that at the end of the partnership, the Messrs. Birley would certainly have the exclusive benefit of all his labours except in this disputed area. Was he asking too much to be allowed to live and let live, and if this could not be universally applied at least it should extend to friends and deserving relations. He finished with the warning that 'My nephew stands in that position and I shall not desert him'.

Trying to conclude with a little financial pressure if conciliation failed, Thomas finally suggested that James had been offered the opportunity to combine with a competitor which would result in a loss to Charles Macintosh & Co. of some £4000.

A few days later, on 30 November, Thomas wrote to James saying that he had seen Mr R.M. and that he gathered from his discussion with him that his statement had been 'well conned over' and that there were signs of a thaw. He added the observation that Mr R.M. had said that it was a long story and then 'in the same breath added 'you have omitted so and so which would of course have made it longer'.

(An uncharacteristic flash of humour for which Thomas was not known or just a statement of the obvious?)

The identity of Mr R.M. has never surfaced, but he would appear to have been someone of perhaps middling authority who had known Thomas for a considerable time and, perhaps, had been involved in the original transfer of the hose business from Charles Macintosh & Co. to James Lyne Hancock.

In the event, nothing was heard from Charles Macintosh & Co. for several months and Thomas and his nephew must have believed things were going their way but, four months later, on 28 March James received the following:

> ...we have carefully looked at the Deed of trust of Mr. T. Hancock and also at our partnership articles. We find that Mr. T. Hancock has no power whatever to grant you a licence to make tubing at all. His reservation merely extends to Hosepipe.

If this letter had properly quoted the Deed of Trust the matter would have been resolved immediately, but Thomas, receiving a copy letter, was beginning to get both angry and frustrated. He responded to James:

> I have just received your letter of this date together with the letter from Chas. Macintosh & Co. to you on the 28th in which they fancy there is a distinction to be drawn between a Pipe and a Tube. I have written to them shewing their mistake and told them I have written to you as follows. *'Messers. Chas. Macintosh & Co. labour under a great mistake—your licence certainly covers the articles in question and you need not therefore make yourself uneasy nor cease to manufacture tubing.'* I will write to them on the subject. I hope we shall hear no more of this strange proceeding, a Hose-Pipe is a flexible Tube. A flexible Tube is a Hose-Pipe—where is the distinction? I am fully prepared to go more into the enquiry if need be.
>
> Your affectionate Uncle, THOS HANCOCK.

The last sentence was prescient indeed for it appeared that someone at the Macintosh factory was not to be put off and within a couple of weeks he refused to vulcanize what was described as a tube and not a hosepipe! Thomas had had enough and decided to give Charles Macintosh & Co. a lesson in the English language that would not be forgotten. Bearing in mind that the research Thomas carried out was done without the aid of the World Wide Web it is a remarkable testament to his education, knowledge and library!

Your letter of the 18th is before me and it proves what I suspected that the absence of precise information on the subject of *'Hose'* has been the cause of the errors into which you have fallen. You say *that 'a Hose Pipe means a particular kind of pipe and not every kind of pipe'*—this nobody will dispute but when you say that *'Hose means a knotted or woven fabric which when united with INDIA RUBBER and formed into a pipe for the conveyance of fluids I conceive to be that particular kind of pipe which reserved in the patent'*, it is evident you have written a definition suited to a particular occasion and not such as to comprehend the universal acceptation of the term. It was impossible for me not to have seen from the first that this attempt to restrict my rightful claim ought never to have been made and I thought on reflection it would have been abandoned but when I found it was persisted in I thought it necessary to take some pains in investigating the origin of the term *'Hose'* as applied to Tubes or Pipes—in default of that to trace back as far as possible the sense in which the term has been generally used amongst men conversant with that subject up to the present day.

Thomas continued by saying that he had devoted nearly a week of his valuable time to this investigation and that his notes were sufficient to enable him to write a little volume on the subject never before perhaps examined. However, he had no wish to do such a thing so perhaps they would consider his annotated notes, beginning with the definition in *Herbert's Dictionary or Encyclopedia*, 1835 and in nearly the same words in *Francis's Dictionary (of Arts and Sciences)* 1842. 'Hose—A term given to a flexible tube for conveying water or other fluids to any required point.'

He then got down to business, quoting documents from 1511, 1588, 1690, 1695, 1775, 1732, 1741, 1732, 1720, 1732, 1779, 1798, 1825, 1807, 1833, 1835, 1830, and, finally,1845, all of which, in some way, refer to the transferring of fluids. The quotes themselves have been omitted here for the sake of brevity, but the materials fall into two distinct categories: there are pipes of metal, wood, glass, pottery, and stone, all of which are rigid or inflexible and another set consisting of pipes of leather, canvas, woollen, India rubber (and cloth and India rubber united), which are pliable or flexible.

It obvious from the extracts that whenever pipes made of metal or some other stiff material are spoken of, the word 'hose' is never used, and when the materials are lithe and pliant either the word 'flexible' is prefixed or the word 'hose' is adopted, and Thomas emphasized that the terms as applied to pipes and tubes are nearly coeval and have been used indiscriminately from 1690 to the present day.

Thomas continued:

> I have before me a list of '*Vanchie's patent woven Hose*' with the prices from one and one quarter inch to eight inches and Hepburn's List of riveted and sewn Leather Hose from one inch to four inches and a sample of sewed Leathern Hose three eights bore.... Perhaps I have now carried the subject in this point of view far enough. When I made the arrangement with Mr H H Birley he was scarcely able to write but he fully understood that I made the reservations for the sake of my nephew who had purchased the Hose business and he wrote down on a scrap of paper (which I still have—and the last I believe I ever got from him) as concisely as possible my wishes. When the deed of trust was agreed on, Mr Brockedon had any [sic] (no?) idea that the covering or enclosing the wire of the Electric Telegraph would be an immense affair (this covering was I think about three eights bore) and then proposed to confine the reserved manufacture to '*Hose Pipes for the conveyance of fluids*' to which I assented. I have now shewn that the word '*Hose*' as applied to Pipes or Tubes has always been used to denote their flexibility. I have also shewn how the words were introduced into our arrangement....

And Thomas continued for a further few pages to hammer home yet more references, even regressing to Roman times in his definitions!

We can, however, stop here since reference to Mr Brockedon and the electric telegraph explains all. Although the Deed of Trust no longer exists it used the expression 'Hose Pipe for the conveyance of fluids', and there is agreement from all parties on this. Here lies the concern of Charles Macintosh & Co. There are no details of the 'tube' which precipitated this confrontation, but, with the advent of the electric telegraph, there is some evidence that the company believed that if it did not make the point absolutely clear at this stage that James was restricted to making tubes/pipes/whatever *only for the conveyance of fluids*, he might be tempted to manufacture rubber tubing for cable insulation, or even start producing the complete sheathed cable, a potentially massive market which they wanted for themselves—only to find rubber almost immediately eclipsed by gutta percha! That their wording of 28 March let the emphasis fall on the difference between 'tube' and 'hosepipe' rather than the conveyance or not of fluids might initially have been accidental but Charles Macintosh & Co. did not feel inclined to explain that, as it might have given Thomas or James ideas that, by Thomas' very agreement to the inclusion of the expression 'conveyance of fluids', clearly he had already discounted.

Thomas was certainly making his point strongly, as he always did, fighting hard for what he believed to be his rights and those of his family. He never showed a sense of humour in his *Narrative* and it is difficult to decide whether this missive—it can hardly be called a letter—shows irony and maybe even wit, or whether it was just his intention to be so exhaustive that there was no room left for further argument.

Thomas had fought, and was still to fight, many battles in court over the years to defend his rights, but this was not one of them! One can only assume that after reading this last missive Charles Macintosh & Co. realized that neither Thomas nor James had their eyes on anything other than the conveyancing of fluids and a peace was agreed. There is evidence of this acceptance in Thomas' *Narrative,* where he writes, initially beginning in 1825,

> ...He (his brother, John Hancock) so devoted a great deal of his attention to the manufacture of tubing, made sometimes solely of sheet-rubber and also by uniting piles of cloth. This gradually led to the introduction of rubber Hose-Pipe...rubber hose and tubing became a staple manufacture <u>and continues to this day</u>.

However, the best evidence for a declared armistice can be found in the *Official Descriptive and Illustrated Catalogue of the Great Exhibition 1851,* one year after the last letter from Thomas Hancock to Charles Macintosh & Co., where we find reference to:

> Hancock, J.L., Goswell Mews, Goswell Road, Manufacturer. 'Vulcanized India-rubber Hose-pipes and various descriptions of India-rubber tubing. Portable India-rubber shower-bath. Hose reel with garden-Hose attached. Inflated India-rubber bed-chair and cushion combined.

Just one year on from the argument, Charles Macintosh & Co. would certainly have refused to vulcanize some of these articles, and certainly banned them from the Great Exhibition if victory had been theirs.

Although the altercation must have soured a number of relationships between the Hancocks and one or more individuals at Charles Macintosh & Co., it certainly increased the family ties between James and Thomas. The one concrete benefit to James was that he continued with the installation of his own masticator and vulcanizing equipment in Goswell Mews and so obtained complete independence from Charles Macintosh & Co., which would enable him to be his own master in the coming years.

10

The Great Exhibition of the Works of Industry of all Nations

The great exhibition of 1851—Stands of Macintosh & Co., J. L. Hancock, the Gutta Percha Co. (now the Bewley business), The West Ham Gutta Percha Co. (Charles' business) and Charles Goodyear. The Macintosh Council Medal and citation. Correspondence between Sir William Hooker and Thomas relating to some of his exhibits being gifted to the Royal Botanic gardens at Kew—An unusual shoe.

Caught in the middle of the argument between his nephew, James Lyne Hancock, and the management of Charles Macintosh & Co., Thomas had little spare time as the second half of the nineteenth century dawned, and, while he certainly would not claim to have better things to do than protect James' interests, he was deep in thought as to how best to display the achievements and ambitions of the Macintosh company in the greatest showcase event ever to have been seen in the British Isles, the Great Exhibition of 1851.

The concept of, and commitment to, the exhibition—more properly to be known as the Great Exhibition of the Works of Industry of all Nations (or just the Crystal Palace)—had been set out in a speech by Prince Albert at The Lord Mayor's Banquet in the City of London in October 1849. It was reported in full in *The Illustrated London News* of 11 October, and two paragraphs in particular must have stirred Thomas, so completely encapsulating, as they did, his ideas and ideals:

> I conceive it to be the duty of every educated person closely to study and watch the time in which he lives; and as far as in him lies, to add his mite of individual exertion to further the accomplishment of what he believes Providence to have ordained...

and

So man is approaching a more complete fulfillment of that great and sacred mission which he has to perform in this world. His reason being created after the image of God, he has to use it to discover the laws by which the Almighty governs His creation, and, by making these laws his standard of action, to conquer nature to his use.

His concluding remarks were an invitation to Thomas to show the world how he had lived up to the principles with which Providence had ordained him:

Gentlemen, the Exhibition of 1851 is to give us a true test and a living picture of the point of development at which the whole of mankind has arrived in this great task, and a new starting point from which all nations will be able to direct their further exertions.

The Exhibition opened on 1 May[1] and triggered a wave of tourism, wonder, and exultation never before seen in the country. The railways had been spreading throughout the land and for the first time it was possible to travel in relative comfort from as far afield as the Scottish lowlands to London in a day—and millions did so—as well as coming from all round the globe. The opening drew 25,000 guests to the great Crystal Palace exhibition hall, who waited for the procession of nine carriages from Buckingham Palace to arrive at the entrance where a red carpet surrounded the platform prepared for Queen Victoria and Prince Albert. The Queen professed herself 'much-moved...by a sensation I shall never forget'. And later wrote 'God bless my dearest Albert' and her love for 'my dear country, which has shown itself so great today'.

In May almost three-quarters of a million people arrived in Hyde Park to see the incredible Crystal Palace and its contents. In June, there were over a million, and so it continued with over 6 million people visiting the exhibition before it closed in October of that year, one-third of the country's population. The organizers used the brilliant marketing ploy of reducing the price of admission as time progressed so that even the poorest could gain entry if they could get to Hyde Park. The came in their droves to see some 13,000 exhibits, and among those were the stands of Thomas Hancock (Charles Macintosh & Co.), Walter and Charles Hancock (The West Ham Gutta Percha Company), and James Lyne Hancock, while John Hancock, one of Thomas' adopted nephews displayed his sculptures *Beatrice*, which had been exhibited the previous year at the Royal Academy,[2] and *Penserosa,* the latter being bought by Prince Albert as a present for the queen. John Hancock was also one of the sculpture commissioners, with his brother-in-law, John

Nunn. There could be no other family with such a presence! Of course the competition was there also. The Gutta Percha Company, Nickels and Keene, and many other rubber and ebonite manufacturing companies. In the North Gallery, which was dedicated to 'Manufactures from Animal and Vegetable Substances' they all competed—with the exception of Charles Goodyear, who chose to be away from his competitors on the second level of the north-east gallery.

Thomas wrote about his experiences in his *Narrative*, and it is obvious that while he was certainly out to impress with the company's product range and versatility of both soft and hard rubber vulcanizates, his feet were firmly on the ground and this was an advertising opportunity not to be missed. Value, practicality, and availability were the keywords from his stand and money was not to be wasted on any form of glorification. True to his beliefs in the importance of education, he displayed samples of natural rubber just as they were received in Goswell Mews from South America and then showed how they were cleaned and processed so that they could be formed into finished articles. He wrote:

> The year 1851 brought with it the memorable Crystal Palace and the *'Great Exhibition of the Works of the Industry of all Nations,'* and we were not slow in availing ourselves of this opportunity of exhibiting such a general collection of rubber manufactures as the world had never before seen; comprising specimens of almost every article to which the substance had been applied, whether adhesive or unadhesive, vulcanized or unvulcanized, possessing elongating elasticity, or rendered rigid by hard vulcanizing, plain, coloured, printed, embossed, moulded portraits, medallions, tablets, stick and umbrella handles, mechanical applications, toys, and various other things made entirely of rubber, and ordinary and coloured solutions were also there, to which must be added some beautiful specimens of rubber produced by the converting process of Mr. Alexander Parkes. Of course, we had also all the well-known Macintosh articles, such as cloaks, capes, of double and single textures, air-beds, pillows, cushions, life preservers, model pontoons, diving dresses, gasbags, &c. &c. We had the pleasure of witnessing the notice taken of our stall by Her Majesty and Prince Albert; the latter of whom took with him a tablet of vulcanized rubber, on which a few lines from Cowper were embossed; they were so appropriate to the national occasion that I am tempted to insert them; they were surrounded with the national arms, and with the rose, shamrock, and thistle, as an ornamental border:
>
> '... The band of commerce was design'd
> T' associate all the branches of mankind;
> And if a boundless plenty be the robe,
> Trade is the golden girdle of the globe.

Wise to promote whatever end he means,
God opens fruitful Nature's various scenes;
Each climate needs what other climes produce,
And offers something to the general use;
No land but listens to the common call,
And in return receives supplies from all.
This genial intercourse and mutual aid
Cheers what, were else an universal shade,
Calls Nature from her ivy-mantled den,
And softens human rock-work into men.
Ingenious Art with her expressive face
Steps forth to fashion and refine the race,
Not only fills Necessity's demand,
But overcharges her capacious hand;
Capricious taste itself can crave no more
Than she supplies from her abounding store:
She strikes out all that luxury can ask,
And gains new vigour at her endless task.
Hers is the spacious arch, the shapely spire,
The painter's pencil, and the poet's lyre,
From her the canvass borrows light and shade,
And verse more lasting, hues that never fade.

COWPER.
May 1st, 1851'

A short distance from the Macintosh Stand was that of Thomas' old company, now under the ownership of his nephew, James Lyne Hancock, displaying those elastic products that had caused so much trouble for Thomas when he had become caught in the middle between James and Charles Macintosh & Co. a year or so ago; 'Vulcanized indiarubber hose-pipes and various descriptions of vulcanized India-rubber tubing, portable India-rubber shower-bath, hose reel with garden hose attached...'. When his uncles visited him they would have passed the stand of Thomas' old employee Christopher Nickels, whom they might once have regarded as a competitor but who now seemed to have settled into a complementary niche, specializing in elastic webs, braids, and cords. They would then have seen, just a few stands away, The Gutta Percha Company displaying a range of products that they felt were, by right, their own. Charles' and Walter's own stand, The West Ham Gutta Percha Company, was only five stands away, and it would have been very obvious to those in the know that the main difference between the two was that the Hancock brothers displayed 'wire coated with gutta percha for electric telegraph purposes', something not mentioned by their competitors even though the break-up of the original company was over the patent rights for the original coating equipment.

And then there was the creation of Charles Goodyear (Fig 10.1(b)). Charles himself (Fig. 10.2) was described[3] as being

> little more than five feet tall, (with) a complexion the color of buckskin and slightly outturned ears. Behind his square jaw a thin line of whiskers fringed his neck.

His gout was causing him problems and he used a walking stick with a sculpted ebonite (or vulcanite as it is also called sometimes) handle (what else?). His display was described as 'costly in character, but very pure in taste'. Goodyear had spent about 30,000 dollars on his stand, as ever, borrowed money—this time from his long-suffering brother-in-law, William DeForest. He had hired architect Stannard Warne to design his display, which was a fantasy world built round 'rooms' of hard rubber or ebonite. Walls and ceilings were coated with ebonite sheets, and pictures on the walls were painted on the same material. Furniture, crockery, trays, ornaments, jewellery, medical instruments were all there—it was a wonderful world of ebonite, the like of which would not be seen again until the Festival of Britain, exactly 100 years later, by which time plastics had entered the domestic scene. The Goodyear display showed the obsession which had dominated Charles since the mid-1830s when rubber first entered his life and was a monument to this new strange material. It was also the presentation of a showman with no thought regarding practicalities and with no hope of recovering his investment through sales or contracts. It was almost as if he did not believe in the future of 'soft' vulcanized rubber, only the hard material. Incidentally, and contrary to popular belief, the American patent for the manufacture of hard rubber or ebonite was taken out not by Charles but by his brother, Nelson Goodyear. To further complicate the Goodyear/Hayward relationship it is Nathaniel Hayward who is identified on his tombstone as the true discoverer of ebonite.

Both Charles Macintosh & Co. and Charles Goodyear were awarded Council Medals, which were only awarded by the Council of Juries to those exhibitors whose offerings displayed 'some important novelty of invention or application' and was specifically *not* to be awarded on

(a)

(b)

Fig. 10.1 (a) part of the Charles Macintosh & Co. stand at the Great Exhibition of 1851. (b) Goodyear's 'Vulcanite Court' at the Great Exhibition of 1851.

Fig. 10.2 A painting of a 'rejuvenated' Charles Goodyear, painted for the Paris exhibition of 1855 on a sheet of ebonite by G. P. A. Healey (the unpaid bill for this work was found in Charles' belongings after his death!).

the basis of 'excellence of production or workmanship alone, however eminent'. It is inexplicable that Charles made no reference to the Exhibition when he wrote 'Gum-Elastic and its varieties with a detailed account of its applications and uses and of the discovery of vulcanization' in 1855.[4] Between Pages 77 and 87 he documents various inscriptions and certificates which he had received over the years, and there is not even a mention of this gold medal there—one of only three awarded to over 500 American entrepreneurs. Perhaps realization had come to him that, while his display was the more spectacular, the items displayed in the Macintosh display were of more use to the world in general.

Did Thomas Hancock and Charles Goodyear meet at the Great Exhibition? Certainly in his autobiography, *Gum Elastic,* Charles comments, writing in the third person:

> The invention patented by Mr. Hancock is the same as that so fully described in this volume as the heating or vulcanizing process described by the writer in 1839. He hopes before long to have the advantage of a personal interview with Messers. Chas. Macintosh and Co. and Mr. Hancock after which he will be better placed to state his views on this subject.

It is possibly this comment, in a book which is often self-pitying, that has goaded so many American writers to cast Thomas in the role of villain. However, it is just not true, as Goodyear himself shows a page or two earlier when he describes his 'Original Specification of Patent, 1844 as Legally Prepared in 1841' where his mixture consists of one part by weight rubber, a quarter to half a part of sulphur and one-half to one part of white lead. Lead oxide was always part of Goodyear's materials, and its absence from Thomas' products confirms that although he picked up the idea of using sulphur from Goodyear, something he readily admitted in his *Narrative*, his process was not the same and not copied. It was only after 1844 that Goodyear conceded that lead was not essential. Goodyear's reasons for not patenting in 1841 are given as (1) no money to do so and (2) his concerns whether sulphur could be replaced by something less unpopular. Two other reasons were also given to do with confidentiality and international patenting law.

Since this is the only mention of Thomas Hancock in Goodyear's book and Thomas makes no direct reference to Goodyear by name there is no published evidence that they ever met formally, neither is there any correspondence recorded between them. However, there is a document in the authors' archives of unknown provenance and authorship, entitled *Notes on the Hancock Family* and dated 1929, which claims:

> In 1843 Charles Goodyear paid his first visit to England and on each subsequent visit he was entertained by Thomas at his home in Stoke Newington. A firm friendship is said to have been made between Goodyear and Hancock, and this would have been unlikely if Thomas had had details of Goodyear's secret and patented it behind his back.

There is no record that Charles was in England in 1843, but the author has problems with dates in other parts of the article and, indeed, this date precedes Thomas' vulcanization patent application. Any meeting between the two at that time would inevitably have involved discussions about the sulphur treatment of rubber devised by Charles Goodyear. The first known visit of Charles to England was for the Great Exhibition in 1850/1 and he was also present in 1853, and this is probably the year the writer intended. The author did not specify who it was who said that a firm friendship developed, so we are left with the tantalizing possibility that they did resolve their differences and become friends.

Fig. 10.3 Certificate confirming the award of a Council Medal to Charles Macintosh & Co.

I hereby certify that Her Majesty's Commissioners upon the Award of the Jurors have presented a Council Medal to
Charles Macintosh & Co.
for manufactures from India rubber,
shewn in the Exhibition

Fig. 10.4 The Council medal.

This extract from the citation for the award (Figs 10.3 and 10.4) was made by Thomas and appears in his *Narrative*:

'COUNCIL MEDAL.'

Charles Macintosh and Co., 73. Aldermanbury (now 3 Cannon Street, West).—The firm of Charles Macintosh and Co. comprises the names of the men who in Europe have made the most useful discoveries in the art of applying caoutchouc to the most varied uses:—the late Mr. Macintosh, who gave his own name to the waterproof garments, and Mr. Thomas Hancock, whose share of merit in the discovery of the vulcanization of india-rubber we have already mentioned. In going through the collection of articles exhibited by this firm, the importance of the uses to which the substance is capable of being applied, especially since the discovery of the process of vulcanization, can be readily appreciated.

The kinds of fabrics with which the garments called Macintoshes are manufactured have always remained the same, but the garments themselves have acquired more lightness and less smell, and the substitution of vulcanized for common caoutchouc insures to them at the present day a permanent suppleness.

The other services which these fabrics are called upon to perform have been greatly multiplied. Their price having become less, they are capable of being applied in lieu of tarpaulings for covering wagons [sic], carriages, &c. The property which they possess of serving to contain water, and which had at first been made available for a well-known therapeutic use, has allowed of their being made into portable baths, which can be rolled up like an ordinary cloth when not in use. The shoes exhibited by Messrs. Macintosh and Co. are made with much care, and with a degree of elegance, which shows that in Europe these articles are but little used except by the more opulent classes.

It is not only in the making of shoes that india-rubber has been called in to supersede leather; the articles exhibited by Messrs. Macintosh and Co. show the use that can be made of it to form pistons of pumps, and how conical valves of india-rubber can be advantageously substituted for leather or metal ones. Sheets of caoutchouc of different colours, either smoothed or worked in relief, are brought in to supersede moulded ornaments in the manufacture of furniture, of ottomans, and in the binding of books.

The use of vulcanized india-rubber to form the piston-valves in steam-engines on the screw principle has greatly contributed to the employment of these novel motive powers, which are destined in some degree to effect a change in navigation, by allowing steam to come in solely as an auxiliary to the wind. The exhibition of Messrs. Macintosh and Co. comprises a valve of this description, which after six months' use has undergone so little alteration that it may be foreseen that these articles possess an almost unlimited durability. The rendering available the impermeability of their fabrics to gas and air has likewise been extended to the air-cushions which have been long used are now added the air-mattresses, so well adapted as beds for travelers [sic] and invalids, boats inflated with air at once portable, and incapable of sinking, and which for life-boat uses and in inland voyages are capable of rendering great service. The collection of Messrs. Macintosh presents some specimens of this kind of great interest. Lieutenant Halkett of the Royal Navy, by making them in several closed compartments, in such a manner that the stuff being pierced through at one point cannot lead as a necessary consequence to the sinking of the boat, has rendered more certain the services which these machines are called upon to perform. The applications of the elasticity of caoutchouc have also been greatly increased. The wheels of carriages have been surrounded by it in such a manner as to prevent the disagreeable noise which they make upon the pavement. Rollers for inking printing types and lithographic stones have been made from it; it is employed for the making of cushions for billiard-tables, and to supersede the use of sacking cords in bedsteads. Advantage has been taken of the elasticity of an India-rubber band, which has a tendency to return to its primitive length when the action of opening a door has elongated it, for the purpose of forming door springs, the use of which is beginning to spread widely.

The substitution of vulcanized india-rubber for metallic springs in the buffers of locomotive engines is one of great service. The masses of vulcanized india-rubber deaden the shock with an ease which may cause this employment of caoutchouc to be considered as one of the most useful to which it has been applied up to the present day. A certain number of the novel applications are due to Messrs. Charles Macintosh and Co. The ability which they have displayed in the manufacture of caoutchouc has afforded to other inventors of these new applications the means of putting their ideas into practice. Without the discovery of vulcanized india-rubber, they could, moreover, never have been carried out. The jury, therefore, in order to recompense the considerable services rendered in the employment of caoutchouc by Messrs. Macintosh, Hancock, and their other partners, have awarded a council medal to the firm of Charles Macintosh and Co.

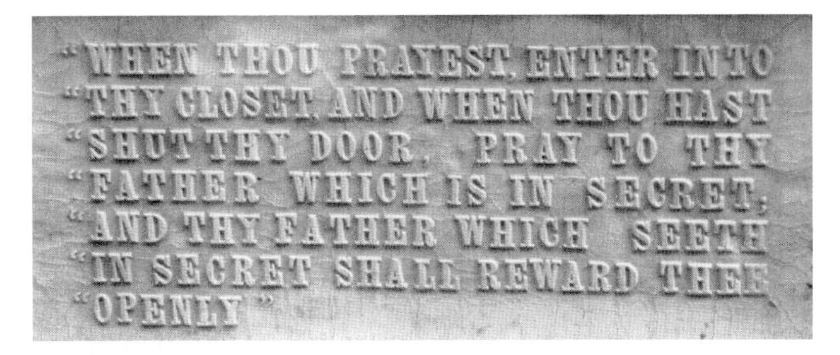

Fig. 10.5 A tablet of hard rubber (ebonite) with raised lettering displayed at the Great Exhibition.

There is no indication of what patterns were raised on these 'sheets worked on in relief', but the obituary of Thomas Hancock in the *Athenaeum* magazine of 15 April 1865 included the comment that his late friend and partner at Charles Macintosh & Co., William Brockedon, 'always spoke of him as having done a service to Science and Art, especially in the matter of elastic moulds for bas-reliefs and he himself always referred with pleasure to his having had the honour to produce a raised type in rubber for blind readers'. The blind Frenchman, Louis Braille, had of course invented his raised letter reading aid not all that long ago in 1829, but there is no evidence that Thomas used that; he was content just to have a raised text on rubber sheets as exemplified by the Cowper poem mentioned above as well as the tablet and plaque illustrated in Figs 10.5 and 10.6.

Soon after the Great Exhibition closed Sir William Hooker, the Director of the Royal Botanical Gardens at Kew, contacted Thomas and asked if he could have some of the products which had been on display to add to the Kew Collection.[5] Thomas must have been distracted by other momentous events, for he completely forgot the request until Hooker reminded him a year or so later!

In January 1853 Thomas realised his omission and, in a letter to the Director (Figs 10.7 and 10.8),[6] we find an apology for 'having so long delayed to send the collection of our manufactures so long promised …' to Kew, while in the middle of the letter we see that the box was due to be sent to Kew on 29 January (the next day) and that Thomas '… thought of running down on Monday (the 31st) to be present when it is opened and give any explanation that may be necessary…'.

Fig. 10.6 An ebonite bas-relief plaque of the type displayed by Charles Macintosh & Co. which could be used to panel an ornamental box.

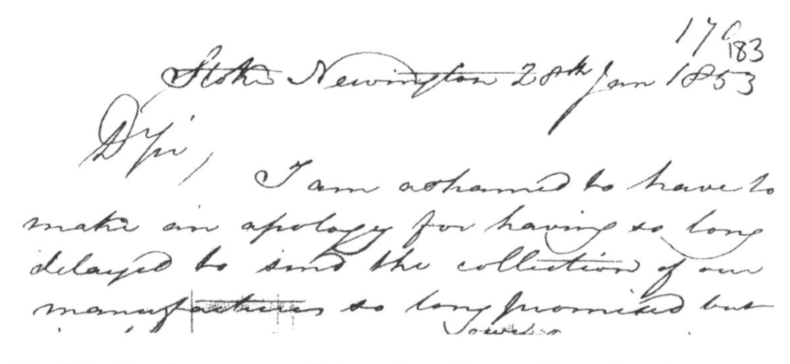

17/183

Stoke Newington 28th Jan 1853

D Sir,

I am ashamed to have to make an apology for having so long delayed to send the collection of our manufactures so long promised but

Fig. 10.7 Opening sentence of a letter from Thomas Hancock to the Director of Kew dated 28 January 1853.

The absence of any reference to Charles Macintosh & Co. or Thomas in the Entry Book[7] on the Saturday while William Brockedon *did* put in an appearance suggests that he acted as the delivery boy. There is no record of Thomas arriving on the Monday and no documents which identify or describe the 40 or so articles in the box.

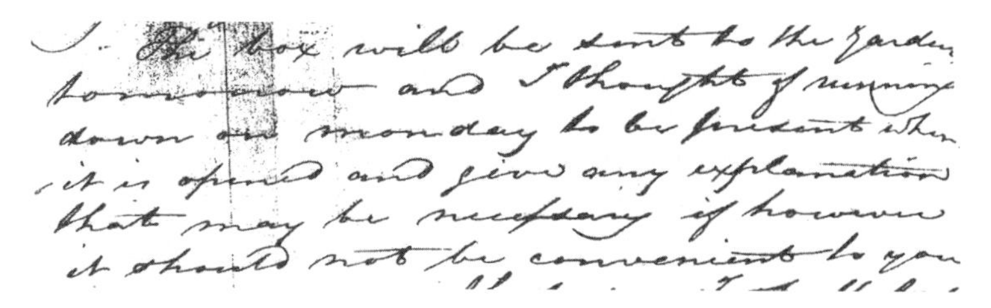

The box will be sent to the Garden
tomorrow and I thought of running
down on monday to be present when
it is opened and give any explanation
that may be necessary if however
it should not be convenient to you

Fig. 10.8 Middle portion of the letter from Thomas Hancock to the Director of Kew dated 28 January 1853.

Fig. 10.9 The mysterious shoe of Kew.

Some 150 years after the Great Exhibition, the box was still in the archives at Kew and it contained many articles which need no description today, bas-reliefs similar to that in Fig. 10.6, a 'gutty' or early golf ball, a child's game of ball and cup, medallions, including one of himself, similar to that in Fig. 7.6 but different in that it had a chip in its edge, and many more, but, of all the items, there was one in particular which stood out—the quite remarkable shoe[8] illustrated in Fig. 10.9. The objects in the box were labelled with paper labels showing their origin, and one on the sole of this shoe confirms that it came from Charles Macintosh & Co. and it appears to be written in the same hand and be of the same age as others in the box.

Thomas' *Narrative* of 1857 (written of course after the donation to Kew) illustrates over 160 soft vulcanized rubber objects and lists many more. Nowhere is there reference to footwear other than overshoes, work boots, or thigh-high boots. If the shoe had been made of leather or fabric its design and style would be perfectly correct for its time to be worn in a lady's home, but it is made of rubber with a complex and delicate embossed pattern which is just visible in a photograph.

Returning to the report of the jury printed earlier we find the comment:

> The shoes exhibited by Messrs. Macintosh and Co. are made with much care, and with a degree of elegance, which shows that in Europe these articles are but little used except by the more opulent classes.

and if we return to Thomas' *Narrative* there are some further significant remarks; he mentions that he made some articles for display using the 'cold cure' process developed by Mr Alexander Parkes

> which consists of immersing the rubber in a solution of the chloride of sulphur in bisulphurate of carbon or coal tar naphtha... the process of Mr Parkes enables us to give to vulcanized articles colours of every tint and a delicately smooth surface... and the most delicate impressions from copper-plate engravings are produced upon them.

This process is especially suited to thin sheets of rubber or articles made therefrom and would be perfect for the vulcanization of an object such as this shoe. This suggests that the single shoe was a unique product that never went into production but was created especially to impress visitors to the Exhibition with the genius of Charles Macintosh & Co. as well as Thomas Hancock.

11

Back to the Courts

Stephen Moulton—background and involvement with Goodyear and friends in America—commencement of a rubber factory in the West Country—patent battle with Thomas—arguments quoted from both sides—inconclusive verdict. Thomas publishes privately a book containing all his patents—American challenge to Thomas' patent—writ of *scire facias* issued against Thomas—his comments on fighting it—final decision in his favour and against Moulton. American realization that Goodyear's patent holds precedent in Scotland—founding of the (American) North British Rubber Company. Thomas suggests plantations to produce clean rubber.

One person who did not have a stand at the Great Exhibition was Stephen Moulton, friend of Charles Goodyear, although he had started a rubber manufacturing business in the West Country in 1848. Because of his importance in the legal battles soon to be fought by Charles Macintosh & Co., it is worth looking at his background in the rubber business.[1] Born in 1794, he had emigrated to America with his wife, Elizabeth, and their nine children sometime before 1842, and there he struck up friendships with a number of rubber pioneers including Goodyear, Hayward, and the three Rider brothers. After trying, unsuccessfully, to persuade Charles Macintosh & Co. to purchase Goodyear's vulcanization process he returned to America and signed a co-partnership memorandum with the Riders and a James Thomas, in which they agreed to share any patent arising from their joint efforts. One interesting observation which Stephen patented in the UK in 1847 was that rubber could be vulcanized with lead hyposulphite. Believing that this would offer a way round Thomas' patent, he decided to open a rubber business in England and chose an abandoned mill in Bradford-on-Avon as his site. There was power aplenty from the river and easy access by road and rail to London. The mill was part of an estate that included

Kingston House, which would become the family residence. Financial backing was offered by the Rider brothers, and restoration of the house coupled with conversion of the mill began immediately. Production began in October 1848.

There is no record of exactly when Thomas became aware of Steven Moulton's factory or how he found out details of his vulcanization process, although, presumably, he would have hot-footed it to the patent office as soon as he became aware that Moulton was claiming to use his own patented process. In the last few years he had had several recourses to law, and he was in no mood to stop now. As we have seen, in 1847 the first major shipment of vulcanized rubber products, mainly rubber over-shoes, arrived in the United Kingdom from the United States. This had the potential to undermine the position of Charles Macintosh & Co. and had to be contested on the strength of Hancock's prior patent. This was found to be valid and after some discussion, Charles Macintosh & Co. granted The Hayward Rubber Company of Connecticut sole rights to import and sell vulcanized rubber footwear in the UK (for a consideration). In a second instance, in 1849, Charles Macintosh & Co. began to prepare a case against a UK importer who was bypassing Hayward, and yet again Hancock's patent was found to stand. With these decisions in his favour, Thomas was content to challenge the patent of Stephen Moulton, but because of all his other commitments he had to wait until 1852.[2]

Moulton claimed[3] that the use of 'lead hydrosulphite and artificial sulphured of lead' did not infringe Hancock's patent, which just used sulphur, or Goodyear's, which used lead oxide and sulphur. He also mixed 'in the dry', whereas Hancock's patent was solely concerned with applications of solutions of rubber (certainly not correct), and there were other differences of varying importance which had allowed the patent to be granted. He further raised the interesting point that Hancock's patent was, in any event, invalid because its deposit paper of 1843 was not followed through in the final specification. The title was:

> For improvements in the preparation or manufacture of caoutchouc in combination with other substances, which preparation or manufacture is suitable for rendering leather cloth and other fabrics waterproof, and to various other purposes for which caoutchouc is employed.

While the text begins with

> ...Preparation or manufacture of caoutchouc in combination with other substances, consists in diminishing or obviating their clammy adhesiveness and

also in diminishing or entirely preventing their tendency to stiffen and harden by cold and become soften or decomposed by heat, grease and oil.

Thomas' trusted solicitor, Mr Karslake, accompanied by Messers Bethell, Rolt, and Webster argued quite simply for an injunction to restrain Moulton from using his process. The claim by Moulton's team boiled down to their assertion that it was the lead which was the crucial ingredient of the process (even though Goodyear had admitted that this was not the case in 1844), while all other matters were ignored. This was demolished when it was shown that the hyposulphite of zinc could equally well effect vulcanization, and in a mathematical simile Thomas' team explained that the plaintiffs had produced a certain result by heating A, B, C, and D while the defendants had used A, B, and Z—but on heating Z it produced C and D! There is little to be gained by repeating all the arguments here as the Vice Chancellor's summing up says it all! It was an interesting, if frustrating judgement for all concerned. The Vice Chancellor found first for the plaintiffs in respect of the patent, which was valid and had been tested in court several times. He also found for them with regard to the chemical arguments which had taken place, being much more convinced of the understanding and chemical know-how of Thomas' experts than he was of Stephen Moulton's. He then tackled the problem of the time lapse and what had occurred during it and whether it was right for the court to interfere by injunction after the delay that had taken place.[3]

In 1849 the plaintiffs, through Mr. Birley, entered into communication with the defendant and asked him what he intended to do. The evidence for the defendant stated that the defendant said that he intended to work according to his patent. This was disputed. On the other side, the statement was that the defendant gave the plaintiffs to understand that he intended to work according to the American basis. The fact, however, was that the defendant had gone on working from the end of the year 1849, and he did not find any direct notice given to him by the plaintiffs that they intended to dispute his right to do so. He did not find that any application was made to the defendant until about March 1852. The reason given by the plaintiffs for this abstinence were that there were actions pending against other parties. He would by no means say that a patentee was bound to bring actions in respect of every infringement; on the contrary the court would not refuse to grant an injunction against any infringer against whom no action had been brought, provided that distinct notice had been given to him that the rights of the patentee would be enforced against him; but in such cases distinct notice should always be given that the party would be held bound by the result of the trial or trials which might be pending. In this case also the actions were brought against the vendors of

American overshoes. The defendant was not dealing with the same actions and if he worked according to his patent, he was carrying on his manufacture by a different process. It was the duty of the plaintiffs to have enquired of the defendant, when he began working in 1849, whether he was working under the American patent and to have informed him, in that case, that they should hold him bound by the result of the pending trial; and if the plaintiffs were not satisfied with the result of their enquiry and had reason to believe that the defendant was proceeding on a different process, they should have applied to this court. On the question of delay therefore he was of the opinion that the plaintiffs had not, as respected the defendant, proceeded with that promptitude which the court required. This case was one of great importance to the parties, and he had considered it with great anxiety and he had come to this conclusion that he must refuse this injunction. The motion was ordered to stand over, with liberty to the plaintiffs to bring an action and to apply to the court in case of any delay in trying the question at law. The defendants were ordered to keep an account, and the parties were to submit to mutual inspection of their works and processes.

As soon as the case was over, Thomas felt it was high time to make sure that those who wished to manufacture rubber goods in England knew exactly what patents he had taken out over the years, so in 1853 he published them all in a book (Fig. 11.1). The full titles and other details of each patent can be found in Appendix I.

The American shoe trade had been considering its position, and was not at all happy with the legal situation as it had been left in England. They saw a potentially large and profitable market if the courts could just be tilted that little bit more in their favour, and a Mr Ross, who was importing American shoes into the UK, but not via Hayward, challenged Charles Macintosh & Co. to sue—which Thomas duly did. After all the old ground was gone over again the jury failed to reach a verdict but the fighting spirit of the anti-Hancock group was high, and in May 1855 they issued a writ of *scire facias* against Thomas, essentially putting the onus on him to provide evidence that he had actually carried out all the work described in his patent and in the time-frame necessary for the patent to be valid.[4]

Thomas provided a graphic description of the trauma which dealing with this writ caused, and it will be heartrendingly familiar to anyone who has stood in the witness box and been cross-examined by an opponent's barrister. It should also be remembered that he had just entered his sixty-ninth year.

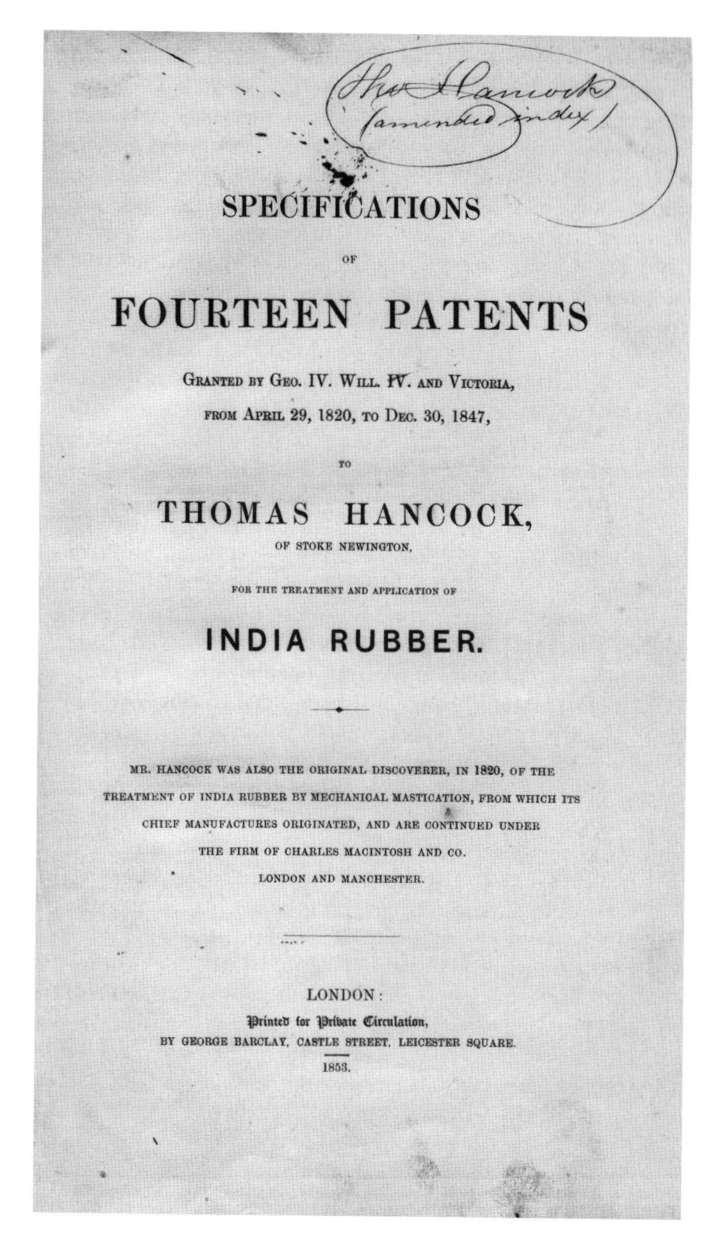
Tho Hancock
(amended index)

SPECIFICATIONS

OF

FOURTEEN PATENTS

GRANTED BY GEO. IV. WILL. IV. AND VICTORIA,

FROM APRIL 29, 1820, TO DEC. 30, 1847,

TO

THOMAS HANCOCK,

OF STOKE NEWINGTON,

FOR THE TREATMENT AND APPLICATION OF

INDIA RUBBER.

MR. HANCOCK WAS ALSO THE ORIGINAL DISCOVERER, IN 1820, OF THE TREATMENT OF INDIA RUBBER BY MECHANICAL MASTICATION, FROM WHICH ITS CHIEF MANUFACTURES ORIGINATED, AND ARE CONTINUED UNDER THE FIRM OF CHARLES MACINTOSH AND CO. LONDON AND MANCHESTER.

LONDON:

Printed for Private Circulation,

BY GEORGE BARCLAY, CASTLE STREET, LEICESTER SQUARE.

1853.

Fig. 11.1 Title page of Thomas Hancock's personal signed 'book of patents'.

If I could have known in 1843 that the laws of the land would have undergone such a sweeping change as that I should be called upon thereafter to give evidence upon oath of all the minute particulars passing in my laboratory during my experiments whilst making this discovery, I might have taken care to have had other corroborating testimony; but I was not possessed of prescience, and had no remedy for the want of it. Supposing I had employed an assistant, there was no provisional specification to protect me in those days, and I might have been betrayed.... I ought to explain that before the commencement of these two last trials the new law of evidence had come into operation; I could therefore be examined on my own behalf, and this doubtless in many cases might be a great privilege, and was so in some measure to me, but it was nevertheless in such a case as mine an anxious and arduous task, and one on which I would not have entered, but that I might have been called on the other side against myself. In ordinary cases all would in general be easy enough; but the reader can have no idea, nor is it possible for me to convey by language the vast difficulty of being prepared (f)or such an ordeal as I had to undergo, nor could the legislature, when passing the new law, ever have contemplated such a case. I will just mention some of the allegations which were preferred against the validity of my patent:—that I was not the first inventor in this realm; that I had not invented that which I claimed; that if I had invented it, I had not invented it at the time I applied for my patent; that there had been a publication of the invention by the exhibition of articles made in the same way, &c. &c.

Now, reader, you may judge of the hardship of my case. How could I prove that I was the first inventor? How could I prove that I had made the invention at the time I applied for my patent? I could assert it, but how could I prove it? Twelve years had elapsed, I had grown old; and my memory was failing. I worked alone; not a human being entered my laboratory during the whole time of making those tedious experiments I have mentioned; I could, therefore, call no witness, I had none. I kept no record, not having the least idea that I should ever have to give any account whatever of these proceedings: how was I to call to mind the minute circumstances? the order? the (s)tages of progress—the moments of favourable appearances, and the substances or circumstances that attended or occasioned them? And yet this was pressed upon me with all the subtility [sic] of one of the most astute and able counsel that could be procured, sifting, catching at every word—perverting or altering my language—urging me to state the minutest particulars during my researches and experiments, in the endeavour to bring to light a discovery, acknowledged—be it remembered when made, to be one of the most abstruse and obscure in science. Under this pressure, I could only have recourse to my failing memory, as to what I did myself; and to such circumstances as I could bring to mind or obtain, corroborative more or less of the facts. I was enabled to grapple with all these proceedings in a manner that surprised myself; at the suggestion of my able and indefatigable solicitor, I ransacked all my drawers, boxes, and receptacles, and amongst my experimental scraps, found plenty of very early ones that were vulcanized; but mostly without dates. There was fortunately, however, one dated the 4th of June, and several in August, one of which was hard vulcanized. Of these dates, I could be quite positive, although they were nearly obliterated; and the scraps were ragged, insignificant little things. Then my servant could prove that ice was taken in daily for my private use, from the ice cart that passed my gate; she could also prove that when my laboratory was pulled down in March, 1843, I used the kitchen oven during the summer till it was rebuilt; I also found the bills for sulphur (with regular dates) procured in the village at the time. These not only assisted my memory, but were in themselves all substantial evidence, and when listened to with attention by an intelligent jury, they readily found a verdict in my favour, and came to me after the trial, and said they had never heard evidence given with more clearness and satisfaction to themselves.

In January 1856 the saga came to an end with the jury finding for Hancock and Moulton having to pay an annual licence to use his own process! There have been suggestions that Moulton was extremely angry with this verdict and maintained a strong dislike of Thomas Hancock for the rest of his life. He certainly felt that somehow he had let Goodyear down, but the only comment about the verdict in a book published privately by George Spencer, Moulton & Co. Ltd.[5] is that:

... he [Moulton] was forced to manufacture under licence in order to continue in business. This was not, however, entirely unfortunate inasmuch as it gave him greater freedom in his choice of materials and processes.

And it was only for a very few years. He had more reason to be irritated by the later actions of Charles Goodyear, who tried to use Moulton's company as his own private and unpaid research and manufacturing base whenever he was in the United Kingdom.

Thomas obviously felt aggrieved that he had been put through this ordeal, mainly as a result of pressure from American manufacturers, and he concluded the biographical part of his *Narrative* with the following observation:

Though called upon to do so, we have never interfered in the disputes on this question abroad, nor have we made any attempt to monopolize any portion of the rubber trade in America, or any other foreign country.

However, this is not quite the end of the story. An American realized that there were two patents by Goodyear and Hancock covering the UK. In England Hancock's had preceded that of Goodyear by some two months, but in Scotland (or North Britain as the country was often

Fig. 11.2 Sir William Jackson Hooker.

called at that time) the position was reversed with Goodyear's application being dated three months ahead of that of Hancock and the final specification just one month ahead. In 1855 the North British Rubber Company was founded in Edinburgh, having been shipped over, lock, stock, barrel, and key workers from the United States.[6] The labour contract between the American rubber workers who were brought to Scotland by the company offers an interesting insight into what was on offer and what would be considered a reasonable rate of pay in the rubber industry at that time.

Typically, a contract stated that 'for the consideration hereinafter mentioned' the labourer would agree to sail to 'Glasgow, Scotland in the Kingdom of Great Britain' and there make such preparations as were necessary for the manufacture of rubber boots and shoes and then continue the manufacturing for a period of one year. He or she would also make him- or her-self generally useful to the company by teaching and instructing others how to carry out the manufacturing processes. In return, the North British Rubber Co. would pay the labourer the sum of one dollar per day from the time he or she arrived at Glasgow through all the setting up, manufacturing, and training period. The company also covenanted to pay all necessary travelling expenses of the labourer to Glasgow from New Brunswick, including the sea passage, and if the labourer wished to return to America at the completion of the contract, the company would pay his or her passage and necessary travelling expenses from Scotland to New Brunswick, New Jersey.

There are many different ways for determining what one dollar per day in 1855 would be 'worth' today but taking basic purchasing power as the most appropriate, it represented around £90 per day in 2007.

With a 'package' of this size and completeness arriving on the shores of Scotland, the opinion of Hancock's legal advisors was that the delay in filing his Scottish patent would probably lead to defeat if he went to court, and since it had so little time left to run the battles finally ceased.

Even during this final court battle Thomas was becoming ever more concerned over the future of the industry he had created, as it was obvious that the demand for rubber was rapidly outstripping supply. He was also aware from practical experience concerning the damage regularly occurring to his equipment that the quality of the rubber entering the country was not as good as it could be if the tapping, collection, and shipping were all more closely under the control of one body. In a letter to the *Gardeners' Chronicle* in 1855, he suggested creating plantations in suitable British colonies such as Jamaica and the East Indies. The idea was taken up by Sir William Hooker (Fig. 11.2), Director of the Royal Botanic Gardens, Kew, who committed himself to render any assistance in his power to parties disposed to make the attempt to move the Amazonian rubber plant from Brazil to territories within the British

Fig. 11.3 Sir Joseph Dalton Hooker.

Empire. Nevertheless, it was to be 20 more years, and at a time when the gardens at Kew were under the directorship of Sir William's son, Sir Joseph Dalton Hooker (Fig. 11.3), before anything came of that commitment.[7] Even so, it cannot be denied that Thomas could justifiably claim to have set in motion the next great phase in the development of his industry.

12

A Life of Ease(?)

Thomas writes his *Personal Narrative* (his life in the rubber business) and distributes it widely. Testimonial to his lawyer, Henry Karslake—correspondence to and from various important figures including Karslake, Sir W. J. Hooker, and Edward Woodcock—Letters to several American 'Rubber Barons' (Day, Buckingham, and Hayward) with copies of his book asking if any similar has been written in America—Buckingham claims not—appreciation of Thomas expressed by all. Thomas retires to be looked after by three of his unmarried nieces at Marlborough Cottage. Presentation of a testimonial from the workers at the Macintosh factory expressing their appreciation of such a great, fair, and generous master—Thomas' reply.

Although Thomas was now in his seventy-first year, he could not rest until he had completed his final *tour de force*—his *Personal Narrative*—which he described as:

...in no respect to be considered a treatise on Caoutchouc; it is simply an account of my progress in the manufacture—feeling confident that no-one has preceded me in this path.

Although humility is hardly a word one of Thomas' business competitors might have chosen to associate with him, it is a word, together with self-effacement, which truly describes his character when, for a few moments, he could escape from fighting for his rights and be among his friends and family. This is clearly illustrated in the first and last paragraphs of his book. He begins his preface:

In writing a personal narrative, it is impossible to escape the very disagreeable necessity of frequently repeating the pronoun I,—my readers must excuse this unavoidable egotism.

and the text ends, as we have already seen, with:

Though called upon to do so, we have never interfered in the disputes on this question abroad, nor have we made any attempt to monopolize any portion of the rubber trade in America, or any other foreign country.

This bears repeating. Thomas had no wish to conquer the world and only asked that he should be able to earn sufficient money to care adequately for his large extended family, since, of all his siblings, he was the only financially successful one.

He finished writing in November 1856 and the book was published in January 1857. The original edition contained his 149-page narrative and a further 128 pages of information about rubber-producing plants round the world, statistical data on rubber exports from the Amazon and their destinations, lists of applications of vulcanized rubber, and the full texts of his 14 patents: a vast fund of information for any historian or economist in this field. A few of the illustrated pages are shown in Figs 12.2–12.5.

Copies were sent to many friends and people in the rubber industry whom he had met or heard of. Some of the letters he sent and the replies he received[1] offer further insights into his character.

However, on 21 November 1856, the date on which Thomas signed the preface to his *Narrative,* he sent to Henry Karslake what he described as a testimonial (Fig. 12.1) but which the Karslake family called 'The Mausoleum'.

Fig 12.1 Testimonial to Henry Karslake from Thomas Hancock.

It was inscribed:

To Henry Karslake Esq[r]
presented by Thomas Hancock
in
Grateful acknowledgement
of
laborious and successful
professional services
conducted with consummate ability
and
zealous warmth and generous feeling of a true friend.
Nov 21st 1856

The bust is that of Karslake and the lower panel shows his favourite hunter and dog. Perhaps surprisingly for Thomas, it is made of conventional materials and contains no rubber, ebonite, or gutta percha, though it was made by Elkingtons for whom Parkes worked!

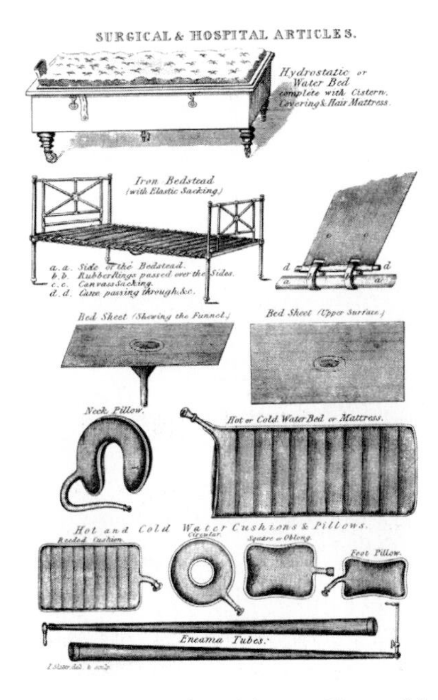

Fig. 12.2 Medical devices—page from Thomas Hancock's *Narrative*.

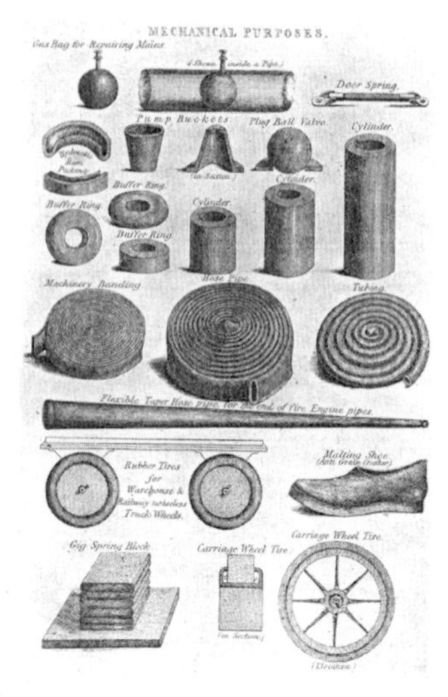

Fig. 12.3 Mechanical articles—page from Thomas Hancock's *Narrative*.

Fig. 12.4 Nautical articles—page from Thomas Hancock's *Narrative*.

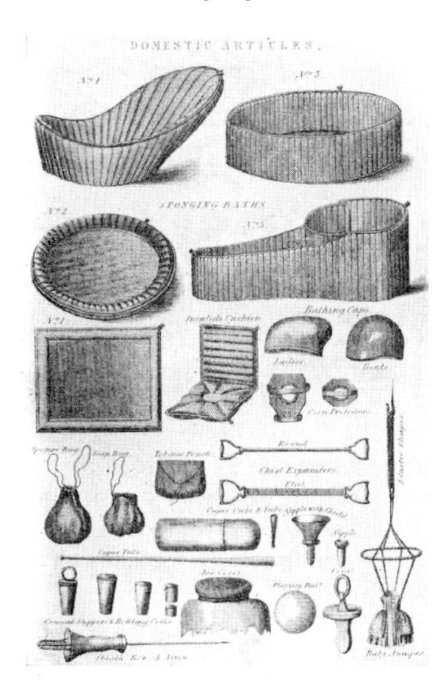

Fig. 12.5 Domestic articles—page from Thomas Hancock's *Narrative*.

A copy of his book could not have accompanied the testimonial as it was not yet printed, but Karslake was very ill at the time and Thomas may have thought that some early sign of his appreciation was due. It took Karslake until the following April to be well enough to reply but he was able then to dictate, for his clerk to write, a six-page reply expressing his appreciation of the gift. He was able to scrawl his salutation at the end of the letter. A fraction of the reply is included here (Fig. 12.6) as an assessment by a man of the law of Thomas' character.

Similar comments of appreciation and affection were received from many writers, ranging from Charles Guibal, whose factory in Paris Thomas set up some 30 years earlier, to a Mr Jas. Miller, who had been lent a copy of the *Narrative* by a friend and who concluded his letter thus:

> I hope, Sir, that you will not think me impertinent in thus presuming to address you, I am a perfect stranger to you, and I do not want anything from you nor am I in any way connected with your trade, but as a humble working man, I feel anxious to record my high opinion of merit and genius, this is the only object I have in view and if you will accept it as such I shall feel proud.

Thomas must have been truly moved by such comments.

and when by degrees I obtained a complete History
of these affairs from the first to the last, that I had
to deal with a Gentleman possessed of the highest
quality of mind for Industry, Perseverance, Truth, and
Integrity, and that you would rather sacrifice an
earthly fortune than deviate in the slightest degree
from the line of Truth and honor.

And when by degrees I obtained a complete History of these affairs from the first to the last, that I had to deal with a gentleman possessed of the highest quality of mind for industry, Perseverance, Truth, and Integrity, and that you would rather sacrifice an earthly fortune than deviate in the slightest degree from the line of Truth and honor.

Fig. 12.6 Part of a letter written by Henry Karslake to Thomas Hancock, dated 6 April 1857.

One of the first people on Thomas' list to receive a copy of the *Narrative* must have been Sir W. J. Hooker, Director of the Royal Botanic Gardens at Kew, who replied very promptly, but perhaps with an ulterior motive:

Royal Gardens Kew. Feb 7th 1857, Saturday

My Dear Sir
I am reading with very great interest the 'Personal Narrative of the origin and progress of the caoutchouc manufacture' which you have been so very kind to send me. The book has the greater charms for me as bringing my old friend Macintosh so much to my recollection, and because we owe to your liberality two cases in our museum filled with articles manufactured of this wonderful substance. I had no idea till now how much the world owed to your patient and laborious and scientific investigations, for so many of the improvements that have been made in the various fabrics! Our collection is still deficient in one striking preparation of caoutchouc, and that, I presume; under the head 'Moulded articles—hard vulcanised' p.174 of your book. Almost the last visit the late Bishop of London paid me, he took a comb out of his pocket and said 'There is a curiosity: it is the best comb I ever had and 'tis made of India rubber! I would give it to you, but it is a gift to me:- but if you go to Leicester Square you may buy such things.' I went but I could not see what I wanted. Can you recommend me to any wholesale shop where I could purchase various articles, boxes etc as well as combs made of this particular state of caoutchouc?

You will perhaps now discontinue your very arduous labours, after having done so much for science and trade. I see you and I were born in the same year and within two months of each other. May health long be continued to you.

Most faithfully yours, W J Hooker

Thomas replied a few days later in his flowing copperplate hand that he knew of no such shop, but it must have been a mistake if no hard vulcanized articles had been sent and he would rectify it immediately.

By the very next day he had received a reply:

Royal Gardens Kew. Feb 12th 1857

My Dear Mr Hancock,

I must correct myself for saying that there was no hard vulcanised India rubber in the beautiful collection you were so good as to present to us. The fact is, it was, in a measure lost and overlooked from the very numerous objects of other kinds and of this, I did not at first even now recognize it, because it is not possessed of the same ebony or very hard look that the comb exhibits. There's a whip or umbrella handle and of a brown colour, and I should not now have recognized it, but for the name written on it. Unquestionably some more would be acceptable and render our collection very complete. The comb you are so good as to send I must be allowed to keep for my own self as an agreeable memento of you and of your kindness and of your achievements.

Most Truly, my dear Mr Hancock, Yours... W. J. Hooker.

Thomas' inclusion of an ebonite comb with his letter was a nice touch! Incidentally, Hooker was wrong in saying that he and Thomas were born in the same year; he (Hooker) was born some 10 months earlier. They did, however, die closely together, with Hooker surviving Thomas by just four months.

Another interesting letter was received from Edward Woodcock, who, as a young man, had accompanied Thomas to France to help set up the Rattier and Guibal factory. He chose to remain there while his brother, Alonzo, stayed in Thomas' employ, rising to factory manager at the works of Charles Macintosh & Co. in Manchester, just two of the workers who spent most of their working lives under Thomas's protecting wing.

...I beg to thank you for the copy with which you have kindly favoured me, I have already perused the same with great interest and shall often refer thereto with renewed pleasure, I am much obliged and feel highly honoured by your kind mention of my humble efforts to promote the prosperity in this country, and most especially so to see you so fully appreciate and do justice to the energies and intelligent efforts of my Dear Brother Alonzo, at home, he indeed has laboured with heart and soul, and I think with some success, for the prosperity of our Industry of which we have both always considered you the Creator.

With the most sincere wishes that you may, Dear Sir, although growing into years, yet live long and happy in the full and peaceable enjoyment of the fruits of your long and arduous labour.

I remain, Dear Sir, Your ancient pupil and Obt. Servt.
E. B. WOODCOCK.

Thomas did not confine his books to Europe and at least three went to America. There is no correspondence suggesting that one went to Charles Goodyear, although he was in England at the time of publication, and if he and Thomas were meeting one could have been passed over by hand. However, as soon as the *Narrative* was published Thomas wrote to Horace Day, who had been responsible for luring Horace Cutler away from Goodyear in the early 1840s. By 1857 he was an important rubber manufacturer who had been as much trouble to Goodyear as all of Thomas' competitors in England added together. Thomas' letter contains a gentle dig or a steely warning depending on your reading:

My vulcanizing patent will expire in November next when your countrymen who seem very anxious to compete with us will have the field opened to them, but I think the whole business will be completely overdone, and like the trade in overshoes ruined by mistaken calculations and bad management. We are fully prepared for whatever may turn up.

But he also wrote:

I have written a Narrative of the part I have taken in the rubber manufacture in England and I send herewith a copy of which I beg your acceptance. If a similar narrative of an authentic character has been published in the United States I should feel very much obliged if you will be so good as to send me a copy of it.

He makes much the same request in a letter to another American manufacturer, W.A. Buckingham, when he asks:

if any authentic narrative of a similar nature has been published in United States I should feel obliged if you would send me a copy.

The final letter in this trio, all of which were written in January 1857, is one written to Nathaniel Hayward, certainly the first person in the story of rubber vulcanization to identify some effect of sulphur on rubber. Here Thomas wrote:

Considering the position you are entitled to take amongst the foremost in the manufacture of India Rubber in the United States I have thought that having written a simple narrative of the part which I have taken in its manipulations in this country a copy would not prove unacceptable to you. I have therefore forwarded one to your address of which I beg your acceptance. If you have not already done the same thing in America no man I should think is more able or more entitled to fulfill such a task.... I indulge the hope that although fast declining into the vale of years I may yet live to see such a production emanating from the press under your hand. With the expression of every good wish to you as a fellow labourer in the same pursuit, I am, Dear Sir....

Why the underlining in the first two letters? The implication must be that Thomas had neither seen nor heard of Charles Goodyear's book *Gum Elastic* or that he did not consider it *authentic.* If Goodyear had received a copy of Thomas' book when the two met, it seems unlikely that he would not have mentioned his own book. Although there were very few copies in existence and none seems to have been completed in terms of filling the blanks and gaps in the original few copies, it had existed since 1853. The emphasis on *authentic* and the wording of his letter to Hayward does suggest that he has seen or heard of a book, which gives him some concerns! There is no record of any of the American industrialists replying with a description of Goodyear's book, while a letter from Buckingham (Figs 12.7 and 12.8) indicates that not all Americans had a low opinion of Thomas.

In July 1858 Thomas was 72 years old. Although he had worked hard all his life and could see around him, both at work and at home, all the rewards for that labour, he felt that the time had come to retire. All his original partners had died, as had his friend William Brockedon and most of his siblings. James was ill and had but one more year to live, although Charles at least was thriving. If, or when, he got bored

Fig. 12.7 Mr W. A. Buckingham.

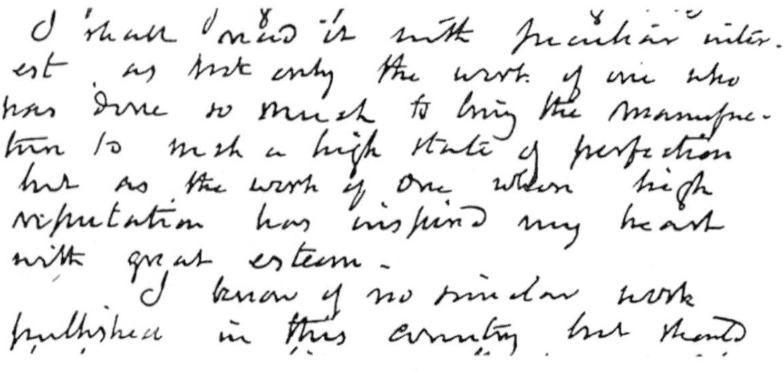
I shall read it with peculiar interest as not only the work of one who has done so much to bring the manufacture to such a high state of perfection but as the work of one whom high reputation has inspired my heart with great esteem.
I know of no similar work published in this country but should

I shall read it with peculiar interest as not only the work of one who has done so much to bring the manufacture to such a high state of perfection but as the work of one whom high reputation has inspired my heart with great esteem.

Fig. 12.8 Part of a letter Mr Buckingham wrote in response to Thomas' letter and book.

he could always go to Goswell Mews and give young James Lyne the benefit of his advice, although whether James, by now a very successful 44 year old, would accept it could be a moot point!

His life at Marlborough Cottage was secure. Three of his adopted and unmarried nieces, Maria, Fanny, and the young, scatterbrained Harriet had continued to live with him, and the 1851 census returns showed that the live-in help consisted of a young maid and a gardener. The gardener's wife was also listed but without a household position. Maria and Fanny between them had cared for their father in Cornwall until his death and then returned to look after their siblings. Aged 41 and 39 respectively they undoubtedly ran Thomas' bachelor household as the 'Ladies of the House'. Harriet was significantly younger at just 34 and had always been Thomas' favourite, coming to live with him when she was just 16 and regarded always as 'the little girl' by the elder sisters.

On 22 July 1858 Thomas was amazed and humbled to receive the following from the Macintosh factory:

Thos Hancock esq.
22nd July

Dear Sir,
It appears that the workpeople profoundly impressed with the sense of your long continuing kindness and regret at your loss as a member of the firm determined to present you with an address embodying their sentiments, the

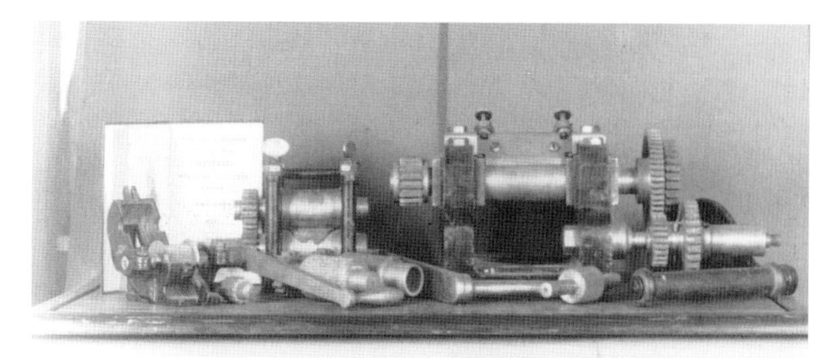

Fig. 12.9 Some of the small machinery used by Thomas in his laboratory in Marlborough Cottage. (His *Narrative*, open at the title page but out of focus, can be identified in the background to show the size of the machinery.)

secrecy observed in a former occasion has been re-enacted and it was not until exhibited to me this morning that I became aware that such an expression was contemplated, or I should have requested permission to add my own signature to a document so identical with my own feelings, however as they will be secret I have to express my great gratification that such a step has been taken, and as requested to advise you that the box will be sent up by rail tonight.

I have to inform you that the address is written out from the draft furnished by the workpeople, I am happy to feel that the sentiments are highly creditable to them and hope that while criticizing the construction of some of the sentences you will kindly regard the faults as evidence of the genuineness of the document.

I now understand that the matter has been long in hand, partly from the delays that usually accompany the proceedings of a body of people and partly a wholesome spirit of competition in the matter of the address.

Trusting you may consider the result in the latter point creditable....

(Unfortunately the rest of the letter is missing, along with the signature). The box duly arrived and Thomas opened it to discover an illustrated address (Fig. 12.10):[2]

Sir
WE, the operatives in the employ of Chas Macintosh and Co., cannot permit the opportunity to pass of your retirement as partner in the above firm, without expressing our heartfelt gratitude for the kindness, generosity and benevolence which you have so liberally bestowed upon us while in your employ.

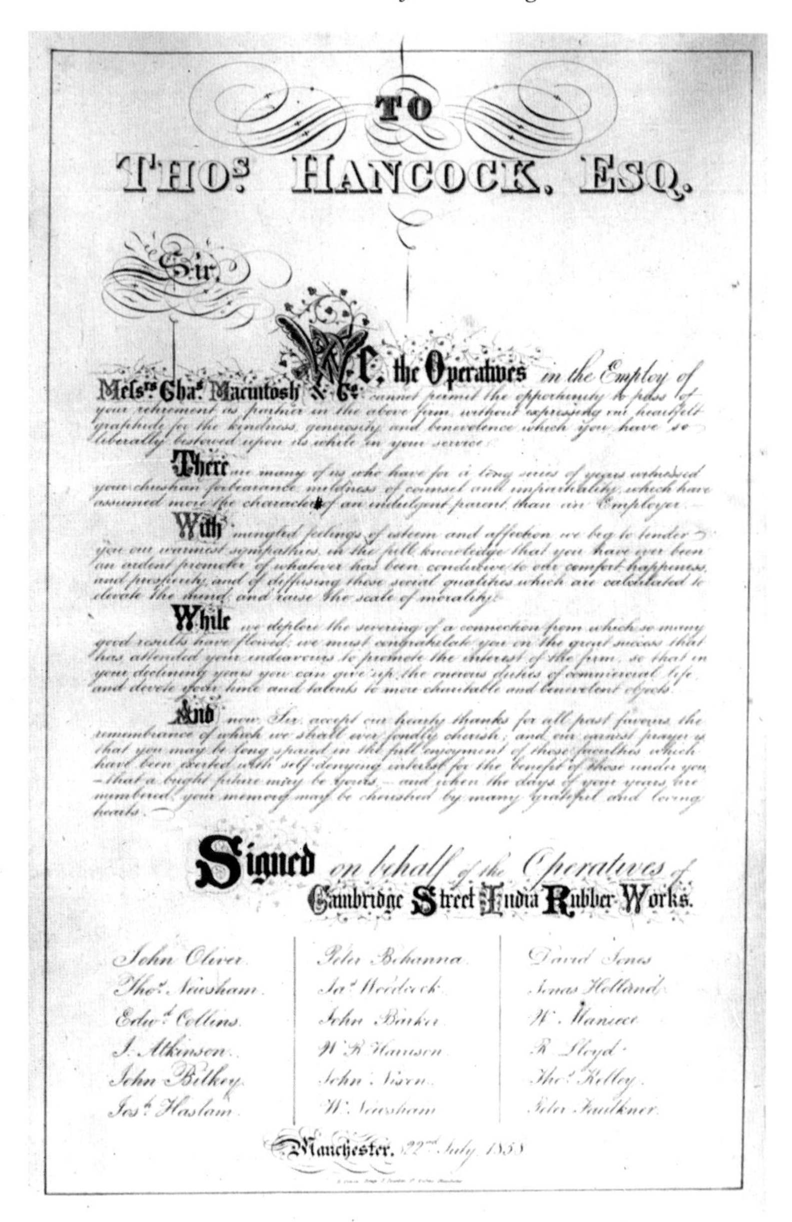

TO

THOS. HANCOCK, ESQ.

Sir,

We, the Operatives in the Employ of Messrs. Chas. Macintosh & Co. cannot permit the opportunity to pass of your retirement as partner in the above firm, without expressing our heartfelt gratitude for the kindness, generosity and benevolence which you have so liberally bestowed upon us while in your service.

There are many of us who have for a long series of years witnessed your christian forbearance, mildness of counsel and impartiality, which have assumed more the character of an indulgent parent than an Employer.

With mingled feelings of esteem and affection we beg to tender you our warmest sympathies, in the full knowledge that you have ever been an ardent promoter of whatever has been conducive to our comfort, happiness, and prosperity, and of diffusing those social qualities which are calculated to elevate the mind and raise the scale of morality.

While we deplore the severing of a connection from which so many good results have flowed, we must congratulate you on the great success that has attended your endeavours to promote the interest of the firm, so that in your declining years you can give up the onerous duties of commercial life and devote your time and talents to more charitable and benevolent objects.

And now Sir, accept our hearty thanks for all past favours, the remembrance of which we shall ever fondly cherish; and our earnest prayer is that you may be long spared in the full enjoyment of those faculties which have been exerted with self-denying interest for the benefit of those under you – that a bright future may be yours – and when the days of your years are numbered, your memory may be cherished by many grateful and loving hearts.

Signed on behalf of the Operatives of Cambridge Street India Rubber Works.

John Oliver	Peter Bohanna	David Jones
Thos. Newsham	Jas. Woodcock	Jonas Holland
Edwd. Collins	John Barker	W. Mansell
J. Atkinson	W. R. Hanson	R. Lloyd
John Bilkey	John Nixon	Thos. Kelley
Josh. Haslam	W. Newsham	Peter Faulkner

Manchester. 22nd July 1855

Fig. 12.10 Testimonial presented to Thomas Hancock by the workers in the Manchester factory.

THERE are many of us who have for a long series of years witnessed your Christian forbearance, mildness of council and impartiality which have assumed more the character of an indulgent parent than an employer.

WITH mingled feelings of esteem and affection we beg to tender you our warmest sympathies in the full knowledge that you have ever been an ardent promoter of whatever has been conductive to the comfort, happiness and prosperity and of diffusing those social qualities which are calculated to elevate the mind and raise the scale of morality.

WHILE we deplore the severing of a connection from which so many good results have flowed we must congratulate you on the great success that has afforded your endeavours to promote the interests of the firm so that in your declining years you can give up the onerous duties of commercial life and devote your time and talents to more charitable and benevolent works.

AND now sir, accept our hearty thanks for all past favours the remembrance of which we shall fondly cherish and our earnest prayer is that you may be long spared in the full enjoyment of those faculties which have been exerted with self-denying interest for the benefit of those under you—that a bright future may be yours—and when the days of your years are remembered, your memory may be cherished by many grateful and loving hearts.

SIGNED on behalf of the operatives of CAMBRIDGE STREET INDIA RUBBER WORKS by John Oliver, Thos. Newsham, Edwd. Collins, I Atkinson, John Bickley, Josh. Haslan, Peter Bohanna, Jas. Woodcock, John Barker, W.R.Harrison, John Nixon, W Newsham, David Jones, Jonas Holland, W. Maniece, R. Lloyd, Thos. Kelly and Peter Faulkner. MANCHESTER July 22nd 1858.

There cannot be many senior executives today who could receive, let alone read, such a document addressed to them with a dry eye!

In Thomas' reply to the workforce, dated 5 August 1858 and addressed through Alonzo Woodcock, the identity of the writer of the letter preceding the testimonial becomes obvious. Thomas wrote:

I have had the pleasure of receiving from you a testimonial recording the expression of the kind feelings you entertain towards me after the long period in which I have been associated with the firm of Messrs. Charles Macintosh & Co., during which time I have with great pleasure witnessed the zeal, industry and good will which you have exhibited in the prosecution of your labours, and the order, subordination and mutual respect for each other in your several departments.

In reading your address, I felt that I could flatter myself that the kind feelings you have expressed should only to some degree be safely appropriated, the satisfaction will be such as to leave an indelible impression during the remainder of my life.

There are two omissions which I regret to find in the subscribing columns, the first is the name of Mr Woodcock, which arose, I understand, from his not

Fig. 12.11 Thomas Hancock, painted by Charles Hancock circa 1857.

being aware of your intentions—I am sorry for the omission, in this instance, because it deprives me of the signature of one amongst you for whom I entertain a very great esteem, and whose long continued, faithful and energetic services I feel it a pleasure to record, To the other omission, I will not give a name, because I am informed it occurs through a recent defection, which ill accords with opinions I had formed of the individual in question, and is a sad return for the liberal treatment which for so many years he had experienced. I turn with pleasure to the honorable contrast which your esteemed address exhibits, and with great satisfaction congratulate you upon your fidelity and your loyal feeling towards your employers.

And now in conclusion, I have to tender you all my hearty thanks for the good wishes you have collectively expressed towards me and my best wishes for a long continuance of that harmony and concord which can alone render you happy and contented whilst in the employ of the firm who (from the intimated knowledge I have of them) desire nothing more than to promote your comfort and welfare whilst engaged in your several occupations, of which with a good number of you with experience of many years, is the best and surest guarantee.

Fig. 12.12 Charles Hancock circa 1870.

I remain with great respect and the kindest remembrance, the sincere well-wisher of every individual among you.

Thos. Hancock, Stoke Newington, Aug 5th 1858.

Through more than 40 years in business, 38 involved with rubber manufacturing, Thomas had maintained his religious convictions and *practised what he preached*. It is obvious that he was considerably hurt by the one defection he refers to in his reply, and the implication is that this was an extremely rare event.

Thomas was a comfortably wealthy man in his retirement. He seems to have *banked* the royalties due to him from Charles Macintosh & Co. with the company because on 20 September 1858 a letter[3] was sent to him showing that he had £28,751.13.6½ deposited with them. It was proposed that this would be paid to him in six-monthly instalments of £5,000 and that the final payment would include all the accumulated interest, calculated at 5 per cent.

13

Death and Dispositions

Death of Thomas Hancock described by his religious mentor, William Benson. Thomas' will—attempted fairness and generosity to family but disputes and involvement of the lawyers—parallel with Dickens' '*Jarndyce* v *Jarndyce*'. Obituary—Benson publishes his *Discourse on Thomas Hancock*. Retrospective (1924) assessment of Thomas and comments on the other family members by Charles Hancock's grandson—Thomas' memorial in Kensal Green cemetery—death of Charles.

In the early months of 1865, when he was 79 years old, Thomas' health suddenly deteriorated. For five or six weeks he struggled with increasing breathing difficulties, and a great terror of death.

His spiritual mentor and friend, William Benson, was later to write:[1]

He was so convinced of sin he could not look for salvation in his own works, and mourned greatly before the Lord, and more and more so to the end when he cried and the Lord did not seem to hear and answer; and in this he did not foolishly justify himself and rebel against God, but acknowledged that it was his own iniquities that separated between him and God, and his sins that hid his face from Him... Notwithstanding his cheerful disposition, there was a weight upon his spirit that made him still a mourner.

In his last illness... in which he had at times felt very near to death, he had been led to cry earnestly to the Lord, and had made many supplications, and had found... a very sweet hope, though not that clear response which he had desired and expected. About the same time he wrote to Mr Gilpin of Hertford: 'The psalmist says it is good for me that I have been afflicted, and I hope I may say so too: bringing untold things daily to my remembrance that had long been forgotten, and with them the considerations of the long suffering and forbearance of God in the midst of sin and follies which now seem almost incredible. This at once humbles me in the dust, and makes me more and more sensible of my lost condition, and the impossibility of being saved in any other way than by the free mercy of God, without worth or worthiness of any kind in the objects of it.'

His niece Maria (pronounced Mariah) recalled[2] that after this, on two occasions especially, he spoke for some time with much feeling of many scriptures and hymns, especially the 103rd Psalm and part of the eighth chapter of Hebrews, 'their sins and their iniquities I will remember no more...'. Once he seemed greatly encouraged when he had prayed for the relief of his breath, and it was granted immediately; then he spoke much of the feeling of his sins, and how unworthy he was of the blessing he was seeking.

As his weakness increased he desired to hear nothing but words of scripture or hymns, saying:

I seem to be able to attend to the Word of God, **that** does not weary me, but I have no strength for anything else.

Yet, about a fortnight before his death, recovering from a bad attack, he said to one of nieces:

I thought you would be frightened, but I was not at all alarmed; I felt ready and willing to go.

She recalled that this was the more remarkable because he had naturally a great dread of pain and death.

Benson wrote that:

When the hour itself approached and his friends were watching round him... he expressed a desire to depart; and when his friend Mr Nunn enquired if Christ was precious to his soul, he answered 'very, very...' After this, he looked round to his eldest niece who had been exceedingly desirous for his spiritual deliverance from all his fears, and said: 'Maria, I feel at liberty;' and when she enquired 'In your soul?' he bowed his head twice in assent. He hardly spoke any more, but soon after gently ceased to breathe, having throughout his illness shown the greatest patience and kindness to all around him.

Thomas Hancock had been ill for some time in 1857 and perhaps it was this illness that prompted him to make his one and only will in August of that year. Nevertheless, he had reviewed it subsequently and there are three extensive codicils dated September 1859, May 1863, and December 1863, the first of which revokes two earlier codicils of February 1858 and June 1859. With his last breath began a burst of activity. His solicitor Robert Wheatly, the surviving partner of Francis Abbott, who had drawn up the will in 1857, was informed of the death, and asked to produce the will to see if Thomas had left any specific directions with regard to his funeral arrangements. Now came a revelation about the manner of the writing of the will. Thomas had called on Mr Abbott and had discussed the disposition of his estate in great detail. The conditions of the will were agreed, but no amounts were entered, blank spaces being left. As Mr Wheatly wrote:[3]

> Thomas Hancock was a well educated man, and most careful and methodical in his habits and mode of conducting business matters, but of a very reserved and reticent disposition and very averse to anyone knowing the state of his affairs or the contents of his will.

Nevertheless, by the time that the will was proved in June 1865 the blanks had been filled in. For such a self-effacing character it came as no surprise that there were no arrangements specified for the disposal of his mortal remains, except it was known that he had booked a plot in Kensal Green Cemetery where so many of his friends and contemporaries were buried and had thoughtfully included the provision of space for other members of his family in the future.

Wheatly was required to read the will in front of the family on the day following Thomas' funeral. He rehearsed the manner in which the will had been made and later amended. There had been two copies of the will made. Mr Abbott had been instructed to keep one copy, sealed within an envelope, which was placed inside another envelope and doubly sealed again. The other copy, which had been similarly treated, was handed to Thomas to place in his own safe in his laboratory. Some years later, when Mr Abbott had been ill, Thomas had taken away the packet containing the will and lodged it with the Bank of England for safe keeping.

To those who had known Thomas and the frugal way in which he had lived, it must have come a great surprise to find that he died possessed of worldly goods to the value of just under £60,000.[4] Inflation can be measured in many ways; in the mid-1860s the average worker in the rubber industry was earning around £1 per week. Today, that figure is in the region of £400–£500, but it includes a substantial improvement in living standards and a more realistic figure for the rate of inflation would be some 80- to 100-fold. This ratio would value Thomas' estate at close to £5,000,000 in terms of today's purchasing power.

Although memory of the first words he had heard Huntington utter ('The Lord recompense thy work, and full reward shall be given thee…') may have inspired him throughout his life, he saw no need for the ostentation shown by the preacher who had to prove that he had *arrived* by travelling in a glass-fronted coach drawn by four horses. Thomas may have been frugal in his own lifestyle but he was more generous to his siblings and adopted nephews and nieces than available documents show. Some evidence for this is given at the beginning of his will,[5] where he wrote:

I acquit exonerate and fully discharge the estates of my brothers William Hancock deceased and Walter Hancock deceased and my brother Charles Hancock and my nephews John and Thomas Hancock and my niece Ann Hancock and each and every of them of and from all debt and debts sum and sums of money which they or any or either of them may owe to me at the time of my decease and all claims and demands whatsoever on account thereof.

He was also a governor of the Wiltshire Society,[6] which existed to assist and apprentice the children of the poor from Wiltshire who were resident in London or in Wiltshire; to lend them money to establish themselves in business, enable them to go to sea, or to attend practical science school. Given Thomas' unswerving commitment to education and apprenticeship as well as the moral strictures imposed on the apprentice during his period of apprenticeship this would have been a cause close to his heart.

In the following paragraph came his first bequest:

I give and bequeath unto my nieces Maria Hancock and Fanny Hancock daughters of my brother John Hancock all my leasehold house called Marlborough Cottage with the garden meadows and appurtenances thereto for all my estate and interest therein. And also all my household furniture plate linen china books pictures wine liquors and other effects therein.

As the will was read it became clear that he had taken great pains to divide his fortune fairly among his friends and relations, each according to his or her needs. He must have spent many months considering the members of his wide family and how best he might provide for them. It was a touching tribute to his warm love for them all, right down to the smallest of his great nieces, although his religious principles led him to exclude Walter's mistress, Rebecca Wendon, and her daughter, Elizabeth Ann, as legatees. His stern rectitude also prevented him from forgiving his brother Charles the money he had lent him in 1863, a total of £5,000, and although Charles wriggled and tried to avoid repaying this debt the executors insisted and eventually, on the sale of Charles' premises in Smithfield to the London Corporation, those proceeds were passed directly to the executors.[7] Considering how shabbily

Charles had treated Thomas over the gutta percha patents, this small implied rebuke was milder than he might have expected, or deserved.

His first codicil confirmed his generosity in that the first part revoked various bequests to his cousins as he had paid them their inheritance in advance of his death, and also he extended the scope of his largesse to any great-nieces and great-nephews, as yet unborn, who may be delivered during his lifetime. The second dealt with Charles' debt of £5,000, while the third dealt with the marriage of one of his nieces and John's daughters, whom Thomas had brought up, to a Mr Peaty. He revoked her legacies totalling £5,000 and instructed the trustees to invest the sum of £3,000 on her behalf:

And shall pay the dividends interest and annual income to arise from the last mentioned trust fund unto my said niece Mary Ann Peaty for her sole and separate use free from the control debts or engagements of her present or any future husband and so that she shall not have power to dispose or deprive herself of the benefit thereof in the way of anticipation And I declare that if my said niece shall at any time or times from disease infirmity or any other cause be incapable in the opinion of my said trustees or trustee of giving a discharge for the said dividends interest and annual income my said trustees or trustee shall pay and apply the same for the benefit of my said niece in such manner as they or he in their or his absolute discretion shall think proper and after the decease of my said niece the said last mentioned trust fund shall sink into and form part of my residuary estate.

Thomas had made clear his views about how his nieces and great-nieces were to hold their inheritances in the main bulk of his will:

I do hereby declare that the respective shares of my said nieces or grand-nieces of and in all and singular the premises in which they may respectively share under or by virtue of this my will shall be for their sole and separate use and benefit respectively and shall not be subject to the debts control dispositions or engagements of any of their present or future husbands.

In the same codicil he noted that bequests to others, including Alonzo Buonaparte Woodcock, should be revoked as he had already paid them their legacies, but he does mention that a gift of £1,000 to his niece Catherine, Mary Ann's elder sister, on her marriage to Mr Bourne should not be deducted from her legacy. Mr Bourne was obviously more favoured than Mr Peaty!

In protecting his female legatees, Thomas was asking for trouble. It was only in 1870 that the first Married Women's Property Act allowed women to keep up to £200 of their earnings and to inherit personal property and small amounts of money; everything else (whether acquired before or after marriage) belonged to their husbands. The 1882 Married Women's Property Act gave married women the same rights over their property as unmarried women, allowing a married woman to retain ownership of property which she might have received as a gift from a parent. It was only with the passing of the 1893 Married Women's Property Act that this process was completed, with married women now having full legal control of all the property of every kind which they owned at marriage or which they acquired after marriage either by inheritance or by their own earnings.

These were the infamous days of the Court of Chancery, whose purpose and duty was to 'do justice' on the basis of the Lord Chancellor's conscience, and conscience has no limits. Thus, as Dickens parodied the Chancery proceedings in *Jarndyce* v *Jarndyce* in Bleak House, proceedings tended to drag on for years:

> Jarndyce v Jarndyce drones on. This scarecrow of a suit has, in the course of time become so complicated that no man alive knows what it means. The parties to it understand it least but it has been observed that no two chancery lawyers can talk about it for five minutes without coming to a total disagreement as to all the premises. Innumerable children have been born into the cause; innumerable old people have died out of it. Scores of persons have deliriously found themselves made parties in Jarndyce v Jarndyce without knowing how or why: whole families have inherited legendary hatreds with the suit. The little plaintiff or defendant who was promised a new rocking horse when Jarndyce v Jarndyce\ should be settled has grown up and possessed himself of a real horse, and trotted away into another world. Fair wards of court have faded into mothers and grandmothers; a long procession of Chancellors has come in and gone out. The legion of bills in the suit have been transformed into mere bills of mortality, there are not three Jarndyces left upon the earth perhaps since old Tom Jarndyce in despair blew his brains out at a coffee house in Chancery Lane, but Jarndyce and Jarndyce still drags its dreary length before the court, perennially hopeless.

Chancery cases provided rich pickings for the legal profession and now, despite all the loving care and shrewd judgements which Thomas had made, his will fell into the hands of the lawyers and the Court of Chancery. There arose a doubt as to his exact intentions in this part and a doubt as to the meaning of this clause and the sum mentioned in that part. Before long each infant mentioned in the will had to have separate representation and the process became so attenuated that further children were indeed born into an inheritance and legatees died without receiving their share. These shares then had to be apportioned between their own legatees and so on it went. If only he had only lived another 10 years, all this would have been avoided by the Judicature Acts of 1873 and 1875. His executors fell out even before the administration of the will began. Thomas Francis Hancock, the youngest of John's sons, had been appointed executor, but when the will was read and he discovered that Tyes Farm, Staplefield, where he had lived and farmed for 14 years as Thomas' tenant was to be sold, he rejected his executorship and renounced his inheritance in the will leaving James Lyne Hancock and Thomas Nunn to sort it all out themselves. Presumably, Thomas Francis was expecting to inherit the farm, but the opening comments in Thomas' will suggest that he had already been borrowing from his uncle to support his farming activities.[8] It was to be many years before the estate was finally closed, but apart from trying to provide for every family member and relation in a manner he thought fairest and best he had ensured the safe succession of the business by passing over the Goswell Mews site to his nephew James Lyne Hancock, now aged 51 years, many years earlier.

However, before the lawyers began their machinations there was time for a retrospective look at the life of Thomas Hancock. On 5 April 1865 *The Mechanics' Magazine* published a long obituary summarizing his life's work. Much was taken from his own *Narrative* but it concluded:

Dying only five years after Goodyear, he has left a name which will be remembered as long as the trade exists, and deserves to be known through all time in the first rank of that brave army of inventors who have helped to render our country great and prosperous by the wonderful advancement of its manufactures.

Mr Hancock was never married, but had adopted the seven daughters and two sons of his late brother John Hancock. He was a great admirer of trees, and in his grounds were several hundred, which he pruned, grafted, and cultivated, and in which he took especial delight. He was exceedingly benevolent, and the blank created by his decease will be rendered the more palpable by his many excellencies.

Also in that April, William Benson published the *Substance of the Discourse on the late Thomas Hancock Esq.* (Fig. 13.1), which ran to some 5000 words. In it, as we have seen earlier Benson describes Thomas' last few weeks and concludes:

Thus we believe our dear friend fell asleep in Jesus; and the Apostle says, 'I would not have you ignorant that such as fall asleep in Jesus shall be raised

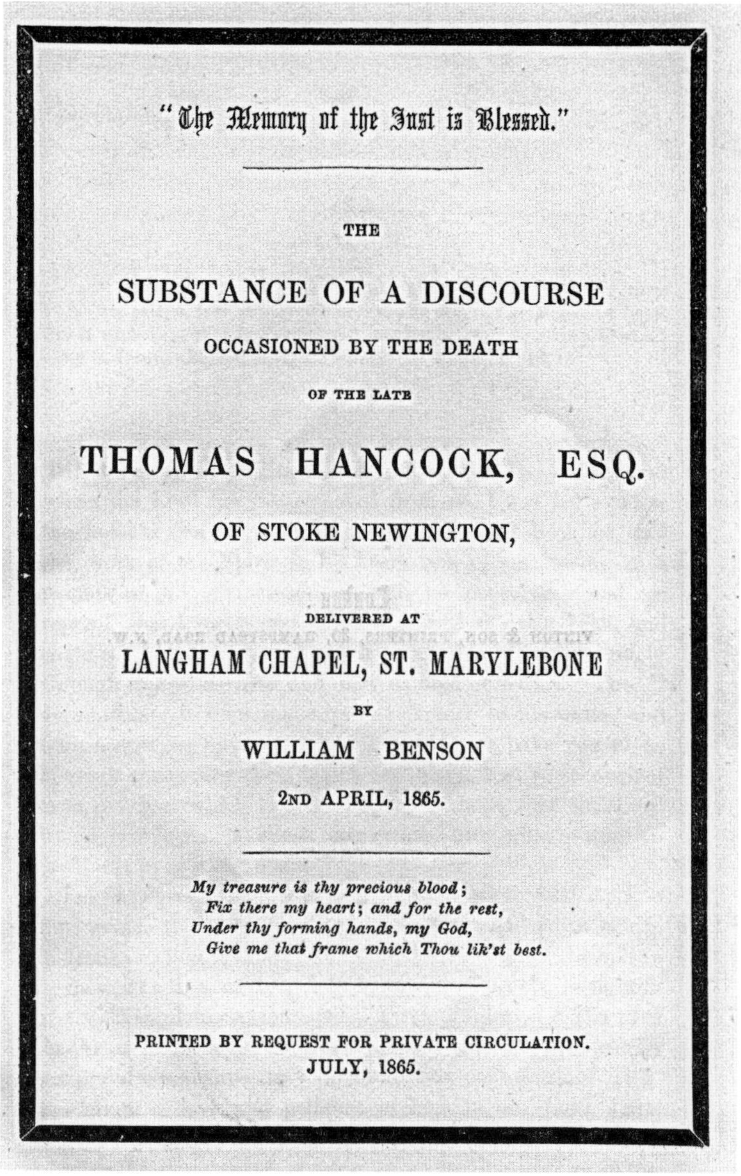

"The Memory of the Just is Blessed."

THE

SUBSTANCE OF A DISCOURSE

OCCASIONED BY THE DEATH

OF THE LATE

THOMAS HANCOCK, ESQ.

OF STOKE NEWINGTON,

DELIVERED AT

LANGHAM CHAPEL, ST. MARYLEBONE

BY

WILLIAM BENSON

2ND APRIL, 1865.

My treasure is thy precious blood;
Fix there my heart; and for the rest,
Under thy forming hands, my God,
Give me that frame which Thou lik'st best.

PRINTED BY REQUEST FOR PRIVATE CIRCULATION.
JULY, 1865.

Fig. 13.1 Cover of William Benson's *Discourse* on Thomas Hancock.

with Him. It is a great and holy hope which the redeemed of the Lord have in Christ, and the value of it far transcends the wealth of this world.

There can be no doubt that Thomas was a deeply religious man and the principles that this commitment carried with it were those by which he lived his life. Benson notes that he was also of 'cheerful disposition' but finds it difficult to allow such a character trait, adding that 'there was a weight upon his spirit that made him still a mourner', almost suggesting that cheerfulness itself was a sin. Certainly, his workforce would not have felt as they did towards him if he only offered them a dour and gloom-laden countenance.

As his obituary pointed out, he outlived Charles Goodyear (memorial; Fig. 13.2) by almost five years and was followed to his grave within four months by Nathaniel Hayward (memorial; Fig. 13.3). The English rubber manufacturing industry was now thriving with over 50 registered companies working in the trade, including the companies of Charles Macintosh & Co., James Lyne Hancock, and Stephen Moulton.

In the 1880s the *India-Rubber and Gutta-Percha and Electrical Trades Journal* published in instalments the text of Thomas' *Narrative* and of his 14 patents relating to rubber. After the final instalment, there appeared on 8 June 1888 a summary of his achievements under the heading 'The Founder of the British Rubber Industry'. The final paragraph read:

It seems to us that Thomas Hancock's place in the commercial history of England is very far from being rightly estimated by the general public. Those who make themselves acquainted with the facts of his career, and ponder over the results of his labours, must allow that he takes no mean position amongst the brilliant band of inventors and discoverers, who, in this Nineteenth Century, have done so much for the maintenance of Britain's industrial and commercial supremacy. And few are the industries which are not in some way aided to or benefited by the use or application of that vulcanized India-rubber, of which the secret was found out and revealed to his countrymen by Thomas Hancock.

This is not a sentiment to be argued with.

Half a century later, in 1924, Messers B. D. Porritt (Director of the Research Association of British Rubber and Tyre Manufacturers) and H. Rodgers (General Manager, James Lyne Hancock Ltd) joined forces to give a lecture on the life and work of Thomas Hancock to the Institution of the Rubber Industry.[9] The Chairman Was Mr Walter Hancock, grandson of Charles and great nephew of Thomas. In opening

Fig. 13.2 Memorial to Charles Goodyear.

Fig. 13.3 Memorial to Nathaniel Hayward—in the shape of the trunk of the 'rubber tree' *Hevea brasiliensis*.

the proceedings, he said it was his duty to offer some explanation of his presence in the chair. It was intended originally that it should be taken by Mr Philip Birley, but, unfortunately he was prevented from being present by illness. Sir Inigo Jones was then asked to preside, but unfortunately his doctor had forbidden him to leave his rooms owing to an attack of bronchitis. Walter was the third choice, which was exceedingly fortunate for posterity since he reminisced for approaching half an hour on the family in general and Thomas in particular.[10] His observations were taken down and provide an assessment only once removed from the lives of the three brothers. Speaking in the third party he reminisced:

With regard to Thomas Hancock, he would not attempt to say anything as to his technical achievements but he could, perhaps, add one or two more or less personal touches which he had gathered from various members of the family in the past. He imagined, from what he had heard of Thomas Hancock, that he was a man of austere character, with a very high sense of duty, and with that peculiarly English trait, an incapacity to show his sympathy and affection toward other people. That he was both tender hearted and affectionate was shown by two facts, namely, that as a bachelor uncle he had adopted, educated and provided for the nine orphaned children left behind in Cornwall by his brother John, and that, although ordinary people never dared to take any liberty with him, a little child could do practically what it liked with him; a child, with its peculiar innate quality of seeing through the external and appreciating beneath the kindness of the heart that beat there.

He remembered seeing, not very long ago, in a review of a book on India rubber—he had not been able to trace the reference again—that Thomas Hancock was referred to as being unscientific. It we took the word 'scientific' in the sense that a man had the capacity for obtaining first-class honours in the Bachelor of Science examination, he would quite admit that in all probability Thomas Hancock was unscientific but, from what had been said that evening, if we took the word 'scientific' in the sense in which it was used by Huxley, that it was organized common sense, he rather fancied his great-uncle must have been entitled to the adjective 'scientific'. . . .

Continuing, the Chairman said he had been interested in one of the portraits shown on the screen more than any other, namely that of his great-uncle, Walter Hancock. He had been under the impression that Walter Hancock was one of those who would never allow himself to be photographed but he expected that it was his grandfather's facetious habit of taking sketches of people when they did not expect it that had enabled us to see a portrait of one who was undoubtedly one of the great pioneers of automobilism. It must be remembered that Walter Hancock was the pioneer of steam traction on common roads and that he had been running a fleet of cars between Paddington and the Bank for about seven months without any mishap. Unfortunately, one of his competitors had met with a serious accident which had resulted in a boiler explosion and the public, not unnaturally, became somewhat alarmed. The aid of the law was invoked and this self-propelled traffic was limited very considerably. It was decreed that the tread at the wheel was to be at least 4 inches, each vehicle was to be preceded by a man carrying a red flag (laughter) and the speed was not to exceed 5 miles per hour. It was probably reduced still further if the man with the flag went to sleep! Those conditions pertained until about 1866 when the Self-Propelled Traffic Association waited upon the late Lord Chaplin, who was then Home Secretary and as a result the law was amended. At the time when Walter Hancock was developing his steam traffic on the common road the attention of the public was being drawn very strongly indeed to the development of railways in this country and that probably detracted to a great extent from the development of steam traction on the common roads. It must be borne in mind that the meeting had been taken back that evening to a time when the locomotive was in its infancy and the electric telegraph had only just been completed by Wheatstone and Cooke, when the internal combustion engine was unknown, or at least undeveloped and it was not until that happened that automobilism as we know it today became the great factor in modern civilization that it has become.

The only one of the three famous brothers whom he could remember was his grandfather, Charles Hancock. One of his earliest recollections was a visit which Charles Hancock had paid to his home when he was five years old, so that his recollection was not very clear. The use of gutta percha as an insulating material for submarine telegraphs and telegraph wire was suggested by Michael Faraday but it was not until the problem of coating thousands of miles of cable with a thin layer of gutta percha had been commercially solved that it had become possible to anihilate distance and to link up, within a second of time, the old world with that new world for which Alexander was said to have sat down and wept.

The meeting had been carried back to an age which was very different from our own and to a time before the brilliant pure science researches of Clark-Maxwell at Cambridge had hinted at the possibility of wireless communication. The history of invention and of discovery was always interesting, and perhaps one of the greatest fruits to be derived from it was the study of the character of the inventor. Coming back to his own family, he said that as far as he had been able to gather, he might say that all three brothers mentioned in the lecture were men who had a high sense of duty, and who had a very great capacity for application and hard work.

It had been pointed out that the early development of India rubber and of gutta percha took place in a period which followed closely upon the devastating influence of the Napoleonic wars. We stand in the world today in an analogous position to that of our forefathers 100 years ago, suffering, as we are,

Fig. 13.4 Memorial to Thomas Hancock in Kensal Green Cemetery.

from the effects of the Great War which closed in 1918. It was difficult for us to imagine where new discoveries and inventions would come from or where they would lead us in the years which lay immediately in front of us which would compare in any way with that enormous advance in technology and in science and in commerce which marked the early portion of the Victorian era. Whether our forefathers 100 years ago were placated with visions of Utopia by suffrage-seeking politicians or not, we could not say. How far the development had been attained was a difficulty. In his own humble opinion the solution of the difficulties which beset industry at the present day—and the rubber industry was no exception—was to be found, not in any drastic reversal of our fiscal system, but in the increased application of scientific methods to our manufacture.

In conclusion he said that not only could the ills from which we suffer be rectified by the application of science to industry but also by hard work, and in that connection he could perhaps not do better than to quote a sentence from the preface of Thomas Hancock's book in which it was said:

'If this book should fall into the hands of a reader who is young, it may serve to stimulate and encourage him to find that, with very slender means at his

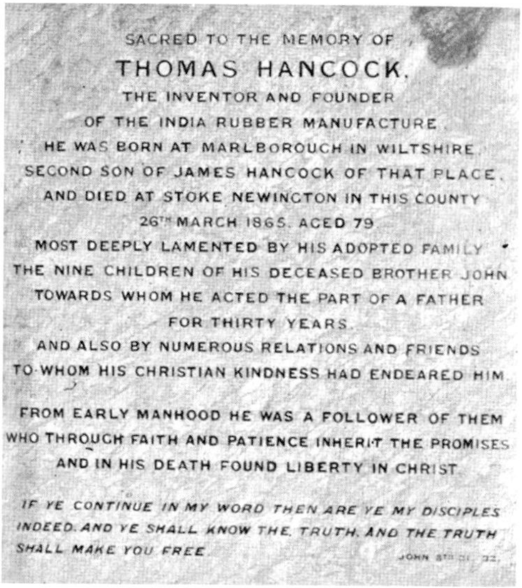

SACRED TO THE MEMORY OF
THOMAS HANCOCK,
THE INVENTOR AND FOUNDER
OF THE INDIA RUBBER MANUFACTURE.
HE WAS BORN AT MARLBOROUGH IN WILTSHIRE.
SECOND SON OF JAMES HANCOCK OF THAT PLACE,
AND DIED AT STOKE NEWINGTON IN THIS COUNTY
26TH MARCH 1865. AGED 79.
MOST DEEPLY LAMENTED BY HIS ADOPTED FAMILY
THE NINE CHILDREN OF HIS DECEASED BROTHER JOHN
TOWARDS WHOM HE ACTED THE PART OF A FATHER
FOR THIRTY YEARS.
AND ALSO BY NUMEROUS RELATIONS AND FRIENDS
TO WHOM HIS CHRISTIAN KINDNESS HAD ENDEARED HIM.

FROM EARLY MANHOOD HE WAS A FOLLOWER OF THEM
WHO THROUGH FAITH AND PATIENCE INHERIT THE PROMISES
AND IN HIS DEATH FOUND LIBERTY IN CHRIST.

IF YE CONTINUE IN MY WORD THEN ARE YE MY DISCIPLES INDEED. AND YE SHALL KNOW THE TRUTH, AND THE TRUTH SHALL MAKE YOU FREE.
JOHN 8TH CH. 32.

Fig. 13.5 The inscription on the plaque at the foot of the memorial.

disposal, and with small beginnings, by care and industry, and with the blessing of God, he may eventually reap the reward of his exertions.'

The last of the brothers, Charles, died in 1877 but little is known of his later years after his involvement with the works of Mr Silver. He was certainly disillusioned and was still being supported by Thomas in his later years as we have seen in Thomas' will. The story now moves to the next generation.

14

Marlborough Cottage and the Great Aunts

J. L. B. James, the grandson of one of Thomas' nieces, Eliza, writes (1950) a memoir of his visits to Marlborough Cottage as a young boy to visit two of the surviving nieces of Thomas (James' great-aunts) who still live there. As a previously unpublished vignette of their life at the turn of the nineteenth and twentieth centuries and their life during the latter third of the nineteenth century, it is reproduced in full—The life and death of Marlborough Cottage after the death of the last of the great-aunts—Recognition of Thomas' place in history by a plaque on the wall of the building currently on the site of Marlborough Cottage.

The Revd James Lyne B. James was the grandson of Eliza James, née Hancock, whose photograph appears as Fig. 16.13. The three ladies in Figs 14.1–14.3 were his great aunts. As we saw in the previous chapter, the two eldest sisters, Maria (1816–1902) and Frances (Fanny) (1818–95) inherited the lease to Marlborough Cottage in Thomas' will and lived there as spinsters until their deaths in 1902 and 1895 respectively. Another sister, Harriet (1824–1909), also lived with them throughout her life. Again unmarried, it is not obvious why Thomas left his property solely to the two older ones; perhaps he thought that Harriet still had a chance of marriage at 34 (at the time the will was written) but that the opportunity had passed Maria and Fanny by at 41 and 39 respectively! It also seems that Harriet was considered the 'little girl' of the family, even though she had two younger sisters, and had never been allowed the responsibilities which usually develop with maturity.

James was a regular visitor to Marlborough Cottage in his youth, and in 1950 wrote this previously unpublished reminiscence[1] of life in

Fig. 14.1 (a) Maria in her youth. (b) Maria painted after Fanny's death.

Fig. 14.2 Fanny in her youth.

Fig. 14.3 (a) Harriet in her youth. (b) Harriet painted after Fanny's death.

the house at the very end of the nineteenth century:

Among my friends none of those who have influenced my life has left a more vivid memory than my three great aunts. They prolonged into this century the life of the early Victorians, both in themselves and in their own gracious home.

To enter their house gave the same sort of sensation that overtakes the tourist when he turns out of the busy Boulevard St. Germain into the quiet residential backwaters in the streets crossing the Rue l'Universite. Everything seems in place but not in time; one seems to have stepped out of one's life into a bygone age. Their house stood in several acres of ground in what had once been the village of Stoke Newington, but which in the course of their long lives had been absorbed into London. While outside their boundaries all was changed, within nothing had changed. There was the rather gloomy looking house of yellow brick blackened by London soot, one wing higher than the other, with an iron verandah running along the lower portion; the front door dignified by two Doric columns, the portico lit by a flickering gas jet.

Behind were ornamental grounds, the lawn paths flanked by formal lines of aloes and leading to a small gate in a wall. On passing through the gate one came to pigsties, fowl houses and a meadow with grazing cows, which stretched down to the New River. To the north of the house, that is to say on the left-hand of the entrance from the road, was another meadow, tenanted by sheep especially brought from Wales and black by nature out of respect for the London atmosphere.

As the crow flies, this domain, modestly called Marlborough Cottage lest it should enter into competition with Marlborough House, could hardly have been more than three miles from St. Paul's, yet it had all the qualities of *rus in urbe*. Incomparably the best way to approach it was from Paddington, as I did so often in my school days. As the train discharged its human cargo the first thing to do was to look out for a sort of kiosk which stood on the main arrival platform—a strange erection, long since swept away, with one of its sides in the form of a multiplication table with ruled spaces, each separate space appearing to be hinged, and when the train was signalled a small flap opened just large enough to reveal a human hand which thereupon inscribed in chalk the number of the platform at which the appropriate train would arrive. Here, at her habitual station, would be found my youngest great aunt, Harriet Hancock, a small slim figure of infinite dignity dressed invariably in black silk.

Even at eighty years of age she had a face of exquisite sweetness and beauty, so striking that those standing by could hardly take their eyes off her, and she was generally to be seen at the centre of a small group gathered round at a reverential distance. She had the long rather sallow face so often depicted by the Pre-Raphaelites, with however a Wellington nose. On her head she wore a small black bonnet crowned by white chiffon, beneath this peeped out her rather straggly gray hair parted in the middle; round her throat was a neatly arranged mass of chiffon tied in a large bow. This contrast of black and white was completed by white gophered frills at the wrists which emphasized the beautiful hands with their long tapering fingers. The coat and dress were, of course, also of black. It could hardly be surprising that the casual crowd watched her with interest since Dante Gabriel Rossetti, a friend of her brother and a frequent visitor at Marlborough Cottage, had been numbered among her admirers and it has often seemed to me that a suspicion of her serene countenance has crept into some of his portraits.

On meeting her, the next thing to do was to look out for her carriage. This presented no difficulty as it also had its own special dignity. It was a large and ancient family coach, so old that in spite of its smart appearance it could hardly have retained any of its original components. The handles of the doors were of solid silver, the windows were lined with vivid blinds of crimson silk, the cushions upholstered in that peculiar shade of grayish-fawn long associated with the first class on a French train [see Fig. 7.1]. To this large brougham there was one special peculiarity, the wheels had solid rubber tyres; this was the first carriage in the world to be so fitted, for the owner's uncle Thomas Hancock had made this invention among many others. The combination of capacity and rubber tyres made the task of the horse a heavy one; this in turn made a peculiarity of the horse, which was always of large build and in colour a sort of piebald. To reach Marlborough Cottage the route lay along Praed Street. Marylebone Road, along the eastern side of Regent's Park, by the Britannia at Camden Town and up the long climb of cobble stones in the endless Camden Road, thence along Seven Sisters Road to the Manor House, with a final right turn down the Green Lanes. The journey was now complete; for from the Manor House, except for a tiny patch at the corner, the grounds of Marlborough Cottage began, an iron fence backed by a privet hedge running down the eastern side until the entrance gates were reached; these were about six feet high, painted in oak-graining and surmounted by a row of metal spikes. At this point a little ceremony took place, something of the kind that occurred in Balzac's famous villa of *Les Jardies* when a creditor called; with the difference that at *Les Jardies* the ritual was designed to keep out the visitor. The coachman deposited his whip in its receptacle and uttered a shrill whistle; this penetrated to the stables and brought out his wife, who thereupon gave warning in the kitchen, whence a maid was dispatched to the sister-aunt. All then went according to plan. The gates were opened, the carriage drove in. As it reached the portico the maid opened the door and in the shelter of the hall stood my elder great-aunt to give a friendly greeting. All was order and comfort, and something in the atmosphere told the newcomer that he entered a truly Christian home—a place where kindness and gentleness reigned supreme and where no evil found place.

My second great aunt, Maria, was the elder and in appearance was not less striking than her sister. Short and rather stout she wore a plain tweed dress, generally of a bluish-gray tint; simple, unostentatious, and dignified, but not as exquisite as the black silk of her younger sister. One glance at the face, however, showed that one was in the presence of a remarkable woman; the face was the face of a homely beauty, the complexion pink and girlish, the eyes kindly, the mouth firm. The silver hair was parted down the centre, dropping two rows of corkscrew curls which recalled the fashion of half-a-century before. These curls appeared to great advantage when aunt Maria went out, for then she wore an early Victorian straw bonnet, specially made at Luton, as the fashion had long since been universally abandoned, tied under her chin with a ribbon of golden satin. Somehow the unfamiliar style of the bonnet made a setting for the face and curls which afforded a fascinating picture.

Inevitably her presence attracted an equal measure of attention; passers-by halted momentarily at the gates as the lumbering carriage drove out, and rubbed their eyes with astonishment on beholding the occupant, showing the sort of surprise that might overtake us if we were to see Queen Victoria suddenly appear in Hyde Park in the costume of a hundred years ago.

The interior of the house had a compelling individuality. Although it seemed so antique it had been in earlier days in advance of its times. For example it enjoyed an ancient but extremely efficient system of central heating, hot air rising through a brass grating in the hall. The windows inside were all strongly shuttered and inside each one hung a handbell suspended on a spring and very sensitive. The doors of the downstairs rooms all had brass bolts at the top, and before going to bed my aunts invariably made a tour of inspection, ending in the securing of the bolts by means of poking them with a parasol ferrule, since they were out of reach. It was never quite clear what the procedure would have been if these two octogenarian ladies had discovered a burglar in the course of the nightly search or what they would have done if the tell-tale alarms had sounded in the still hours of the night, for there was no man in the house, the coachman living over the stables. But of this I am quite sure neither of the ladies would have shown the slightest fear, and no burglar would have escaped without Aunt Maria's rebuke.

Four rooms opened out of the hall, on the right was the dining-room, the windows shaded by the verandah; opposite the front door was a small parlour, never used but full of ancient curiosities which were fascinating to a great-nephew. To the left of the front door were the *'iron-door room'* [Fig. 14.4] and the drawing room. The former gained its name from the corrugated metal door painted in oak-graining. This room was of some interest, it was the laboratory of their uncle, Thomas Hancock, the friend of Faraday and the partner of Charles Macintosh: it was in this room that he discovered the process of

Fig. 14.4 Thomas Hancock's laboratory in Marlborough Cottage (taken from the 1920 reprint of Thomas' *Narrative*).

vulcanizing India-rubber. It remained exactly as it had been at the time of his death in 1865. From this room a private staircase ascended to his bedroom, and as the idea for the crucial experiment occurred to him while he was in bed, he came down at once, made the experiment successfully, and thus came about one of the great discoveries of the world.

The drawing-room, seldom used, for the family, like many of the Victorians, habitually sat in the dining-room, was as representative of the past as aunt Maria herself. I do not suppose that anything in it had been changed since the death of her uncle. Three French windows gave light to the room, opening out into the grounds. In the centre stood an oval shaped table of walnut, covered by a cloth; on this stood an ancient stereoscope, a copy of John Hancock's bust of Milton's *'Il Penserosa'* (sic), and a book given to aunt Harriet by Dante Gabriel Rossetti; incidentally he gave her other things including a brass button from the British sailor's uniform which made his father's disguise when he escaped from Italy. The chairs were of mahogany, all stiff and formal, covered in red damask; before the fire stood a tapestry screen. There were several small tables crowded with knickknacks in the true Victorian style. Upon a polished table was an item of immense interest to all boys, a scale model of a steamship, protected from profane hands by a glass case; above this hung a large oil painting of Cumberland cattle by their uncle Charles Hancock [Fig. 14.5] the animal painter. Altogether the room had the appearance of the familiar picture of Dante Gabriel Rossetti in his drawing-room in Cheyney Row. One

Fig. 14.5 Cumberland cattle painted by Charles Hancock in 1846.

special characteristic I recall was the stuffy smell, as if the windows were never opened.

But the principal sitting-room, as already said, was the dining-room, a long low room with a white marble fireplace, surmounted by a gilt framed mirror reaching to the ceiling. On the mantel-piece reposed two small urns of Worcester china with a carriage clock in the centre. A mahogany book case stood to the right, with drawers in the lower half. Next to this, with its back to the window, was a tall chair without arms, rather gauntly upright, covered in seat and back in embossed royal blue velvet; this was aunt Maria's special chair, although it can scarcely be called an easy-chair. On the other side of the fire beside a brass coal box was Aunt Harriet's chair, a Victorian arm-chair upholstered in red velvet as were all the other chairs in the room. The most obvious piece of furniture was of course a large mahogany dining-table with turned legs, overspread by a stuff cloth.

At the far end of the room, near the door was a large and characteristic mahogany side board, hanging over it a group of horses and other animals painted by Charles Hancock at their country place at Tyes Farm, Staplefield. The side opposite the windows was recessed, with a red plush sofa below and a large picture of Bolton Abbey above. On one end of this sofa always reposed a pile of recent issues of *The Illustrated London News*—a happy quarry for all juvenile visitors.

The hours were regular, breakfast at nine, lunch at two, tea at about five, supper at nine. After lunch when the cloth was cleared the maid put on the table a small walnut and ebony writing desk. Aunt Maria then attended to her correspondence, dealing with each item on the day on which it arrived. Business letters received due attention, bills were paid, family matters dealt with, and finally every charitable appeal received earnest consideration. But if there was much charity there was also an absence of sentimentality and a homely wit. Thus one nephew, addicted to the theatre, received a rebuke. His aunt told him that she was surprised that he spent so much on the theatres, 'the only thing to recommend them is their candour, for they do write outside in large letters: "This way to the PIT". A missionary society unwise enough to complain that people gave so little to missions and left such large sums to their relations got an unexpectedly stinging response. 'I suppose,' wrote this most generous of women, 'that even missionaries sometimes read the Bible; when they do they will find it stated that a man's first duty is to those of his own household.'

A strong religious atmosphere pervaded the house. No attempt was made to be improving, but the least suspicion of want of charity, irreverence, or intemperate language met with a silence expressive of a frigid 'we are not amused.' In a manner wholly natural and cheerful every thought, word and deed took place on the background of personal religious faith. Every day brought a contingent of visitors, ranging from the numerous relatives and dependants to clergymen with needy parishes; no deserving cause failed to enlist practical help and the most boisterous of the great-nephews never left the house without some monetary gift. On the days when the old ladies were going out the brougham came round about 3 p.m. and the remaining correspondence was left over until after tea. Occasionally an afternoon call was paid, but more frequently the excursion was to a philanthropic gathering or to buy necessaries for the objects of their charities; everything had a purpose. Visitors had the carriage placed at their disposal, and were always met at the railway terminus and returned thither. There were, however, certain restrictions. If, for example, the station was London Bridge the occupants were discharged at Moorgate Street, and instructed to take a cab, the journey across the crowded bridge being considered too dangerous for the well-being of the carriage; if it carried by any chance a younger member of the family, the cab fare was always paid for him. Visits to the Bank, Aunt Maria always made in person, the younger sister Harriet being held to be too young and inexperienced to undertake so solemn a mission; although when at the age of eighty she succeeded her sister (the intermediate sister Fanny having died long before) she proved herself a most admirable woman of business.

Supper came on at 9 p.m. and was followed by family prayers. At this gathering of the household Aunt Maria officiated, sitting at the table. Aunt Harriet occupied her armchair by the fireside, with members of the family gathered towards that end of the room and the maids modestly taking the chairs just inside the door. The devotions began with a fairly long passage from the Bible, admirably read; all except the reader then kneeled down with their faces towards the back of their chairs. A long prayer followed, often extending to two or three pages, taken from an ancient book of devotion of a deeply spiritual tone. All this produced a certain religious tension, which on one occasion at least proved too much for the piety of the participants. The cat—which was always a large sandy one, chosen so that 'one does not sit on it by mistake'—having secreted itself in the room unfortunately took upon itself to play leapfrog on the maids' backs. No notice was taken by the reader. A slight tittering escaped from the innocent victims. Still no notice. Two of the great-nephews, not daring to move, began peering through their fingers, twisting their eyes sideways in a dangerous squint. This irreverence brought nemesis. Unable to control themselves they burst out laughing. Yet even at this final crisis not the slightest notice was taken, not a tremor came to the reader's voice. Moreover when the devotions ended and the maids went out not a word was said. To one at least of the participants that silent rebuke cut more deeply than any words could have done. The incident was simply characteristic of that truly Christian household.

In these two elderly ladies the whole Victorian period seemed embodied. Maria was born in 1816 and Harriet in 1824. Their memories went back to seeing Queen Victoria as a child playing in Kensington Gardens; they had witnessed the coronation procession of 1838; the Duke of Wellington had ridden in their carriage. In science they had known Faraday (a Fellow of the Royal Society) and Macintosh; in art and literature they had been associated with the Pre-Raphaelites. They could remember an England without railways and telegraphs; they had travelled from Cornwall to London hatched down in a small cockle-shell of a steamboat and from London to the Lake District in a coach.

Yet if they were ancient they were modern too. Before 1833 they had ridden on the common roads in a mechanically-propelled vehicle, their uncle Walter Hancock having invented the most successful of the early steam coaches. All the beneficent novelties brought to our civilization by the use of India-rubber—save the rubbing out of pencil marks and waterproofing material, which came from Charles Macintosh, their uncle's partner—had first been introduced in their own home. They themselves earned certificates as contributors to the Great Exhibition. At the Great Exhibition they saw their brother John's sculptures [Fig. 14.6], and it has been said that had he lived he would have been the greatest sculptor England ever produced. If they visited the Academy they went to admire the paintings of their uncle Charles Hancock. The Victorian age was a great one, and the Queen who gave the age its name

Fig. 14.6 John Hancock (seated) photographed in his studio in about 1851. The lady is believed to be one of his sisters.

Fig. 14.7 PHS plaque on the wall of Banstead Court.

was a great woman. Among the most loyal of her subjects were the two women at Marlborough Cottage.

Maria lived one year after Queen Victoria had breathed her last, Harriet, a patient of that great physician Sir Richard Douglas Powell in whose presence the old Queen died, survived until 1909, almost to the reign of George V. She was then reunited with her sister and her uncle Thomas Hancock in the shadow of the great obelisk [Fig. 13.4] which marked their grave in Kensal Green Cemetery.

At her death Marlborough Cottage ceased to be inhabited. A strange fate awaited it. The grounds were used for some years as a sports club; a church was built on a portion of the site. In 1944 a V bomb fell in the adjoining property, shaking the old house which had outlived its inhabitants. I visited it in the spring of 1945. The entrance gates had been taken down that very morning; the shell of the familiar building remained, but little more. I went inside. Already the floor boards had been stripped; the whole place looked incredibly mean and the rooms painfully small. Only one room remained intact—the iron-door room* in which the memorable discovery of vulcanization was made by Thomas Hancock; the name of the process as well as the process itself is due to him. This room was temporarily spared for the use of the demolition workers; some of them were eating their lunch as I entered. A friendly workman took me round, telling me that the site was being cleared for L.C.C.

* Although the author mentions the iron door which protected Thomas' laboratory—or perhaps was intended to protect the rest of the house if one of his experiments went seriously awry—he omits to mention that as he was leaving the remains of the property the workmen gave him (or perhaps he just picked up) the lock to the iron door. He also acquired a small piece of the V2 rocket which led to the demolition. Both of these items remain in the family to this day.

blocks of flats. It is a fitting end and somehow I like to think that the site, once the home of these gracious spirits is now given up to the praise or God and the happiness of men.

Today the block of flats is known as Banstead Court. On 8 October 2003 the Plastics Historical Society placed a plaque on the wall by the entrance to celebrate the life and work of Thomas Hancock (Fig. 14.7). It was unveiled by one of the four great-great-great-nephews and nieces of Thomas who attended the ceremony.[2]

15

The Hancock Legacy

Charles Macintosh & Co. after Thomas—the Royal Jubilee celebration in Manchester (1887) and Macintosh's contribution—the problem with odour from rubberized fabric—Dunlop's pneumatic tyre patent—Macintosh's patent battle with the North British Rubber Co. and Bartlett over the 'Clincher' tyre design—Macintosh defeated. Introduction of new chemicals and processing machinery as well as plantation rubber—take-over by Dunlop Rubber Co. of Charles Macintosh & Co. Moulton's rubber business in the West Country—development—relationship with George Spencer & Co.—history through to Moulton's great grandson, Sir Alex Moulton.

Although there were many firms manufacturing rubber products in the UK by the time of Thomas' death, a substantial number were only concerned with the manufacture of waterproof garments and could well have been buying in the treated fabric rather than making it themselves. The direct influence of Thomas was present on three of them: Charles Macintosh & Co. and James Lyne Hancock directly, and Stephen Moulton's factory in Bradford-on-Avon indirectly. Within a few years, however, as patents expired and entrepreneurs saw the demand for vulcanized rubber products grow beyond most people's expectation, new companies sprang up by the dozen and new patents were taken out daily. Life for the old companies entered a new phase.

The hard work of building up Charles Macintosh & Co. into a successful business able to manufacture a wide range of rubber-based products had been completed by the time that Thomas retired. New products were forever being requested and technological improvements led to the introduction of new plant and machinery. Nevertheless, the basic operating philosophy had been firmly established by Charles Macintosh and Thomas Hancock, and since it was obviously producing good returns on the partners' investments there was no need for any radical changes. However, given that the Birley family had insisted in 1824 that Charles Macintosh's original factory in Manchester should be designed in such a way that it could be converted to cotton spinning if the market for rubberized cloth disappeared, it is ironic that during the late 1860s there was a decline in cotton spinning in the area, in part due to the difficulty in obtaining supplies of cotton from the United States, which was caught up in its Civil War, and some of the Birley mills were converted to rubber production.

The continuing success of the company was, no doubt, a matter of contentment for the executors of Thomas' estate since he had willed that:

> And wherein as under and by virtue of the Articles of Partnership entered into by me with my partners constituting the firm of Charles Macintosh and Company and of an Indenture bearing date on or about the eighteenth day of August one thousand eight hundred and fifty three made between Thomas Hornby Birley Hugh Birley and Herbert Birley of the first part me the said Thomas Hancock of the second part and William Brockedon deceased of the third part I am empowered and authorized by my last will and testament to nominate and appoint any person or persons who shall be at liberty and entitled to take my share of the said Partnership business in case of my death during the continuance of the said Partnership business...Now I do hereby direct nominate and appoint the said Thomas Hancock of Tyes James Lyne Hancock and Thomas Nunn to take my share of and in the said Partnership business and...hold the same and all profit and benefit...to fall into and form part of my personal estate and effects and to be applied and disposed of by them accordingly

It was in the interests of the Hancock family that Charles Macintosh & Co. continued to prosper and it did.

There was a significant change in the law in 1867 when the Factory Acts Extension Act and the Workshop Act were passed by Parliament. This brought the rubber industry within the scope of factory and workshop legislation thus prohibiting the employment of a child in a rubber mill under the age of eight (by 1874) rising to 10 in 1876. The working hours of women and young persons were limited to 10½ hours per day. There were potential problems in the last part of this legislation since, as in so many businesses in the nineteenth century, the working day continued until that day's quota had been filled, and this was balanced out by short-time working if the load was low. As the cry goes up today, it was just one more piece of legislation which might add to costs but had to be complied with.

Fig. 15.1 A contemporary engraving of the Macintosh site in 1857.

The drift from a rural to industrial economy throughout the country coupled with the success of the Great Exhibition in 1851 led to a proliferation of what today are known as Trade Fairs. In 1887 there was the Royal Jubilee Exhibition, which was held at Old Trafford, Manchester, at which Charles Macintosh & Co. exhibited. A report of the time[1] describes the stand, which seemed still to be steeped in Thomas' philosophy that one could not have too much education and it is pleasing to note that Thomas himself was still considered worthy of mention:

As might be expected, this historic firm has taken care to exhibit a show of rubber goods worthy of the occasion. Manchester is commemorating, with its Jubilee Exhibition, the fifty years of Queen Victoria's reign. All through those fifty years, and for ten years previously, the Macintosh firm has been manufacturing its rubber goods at Manchester, and sending them out to all parts of the world. Their exhibit is at once attractive and instructive; it is more of an exposition of rubber and its capabilities, and less of a mere advertising display than is often the case in exhibition stalls. Towards the centre the eye rests on a portion of a rubber tree (*Hevea brasiliensis*) about ten or twelve feet in length-a mere fragment of some tall tree that once towered amidst the wild luxuriance of Brazilian forests. A thin section of a similar tree is shown close by. Then we have various blocks of rubber as imported from Para, also specimens of *Mangabeira*, &c. Here, too, are some specimens of the nuts of the Urucari palm, long used by the natives in the preparation of rubber, From being dried over a fire of these nuts the well-known bottle-rubber of commerce acquired a rich brown colour, and a peculiar cooked-like odour.

The exhibit of Ceara rubber (*Manihot glaziovii*) resembles a tangled mass of seaweed. This rubber when it exudes from the trees, tapped for the purpose, is allowed to congeal in rough strings, which are afterwards rolled up into loose balls, or packed in bags. A very fine long specimen of washed Para rubber shows the raw material at a more advanced stage of treatment. A pyramidal display of glass bottles containing various coloured powders, &c., shows us some of the drugs and chemicals used in the manipulation of rubber. The specimens shown give some idea of the complexity of the processes necessary to fit India-rubber for its commercial uses. But the mixing room is an abode of mystery, and every manufactory has its own secrets in connection with it. Rubber in a still more advanced stage is shown by the two solid masses, one red and one black. Here the rubber has not only been washed and crushed and mixed, but has also passed through that great initial invention of Thomas Hancock's—the Masticator-which made the development of rubber manufactures possible. These blocks are in readiness for the cutting machine, and close by are samples of thin sheets cut from them. These sheets are of course for the manufacture of pouches, and various similar articles, also for cutting into threads. Of the latter a large selection is shown in various sizes.

Rubber for most mechanical purposes is treated in a different way. It is made into a dough, with various ingredients, and passed through calenders to form sheets, from which the required articles are shaped. Of these articles numerous specimens are shown, as well as samples of sheet rubber from the calenders—some of it very thin, and some an inch in thickness. Here, too, are washers, valves, and buffers, and all the well known applications of rubber for mechanical uses. We should also name the belting, wheel tires, hot water cushions, the vulcanized rubber for mats, and floor covers of various designs, and the hose (with fittings). The vulcanized India-rubber and canvas garden hose demands special notice. Finding that there is a growing demand for a really good hose to withstand a high pressure of water, Messrs. Charles Macintosh & Co. have introduced a special quality, which is tested, before leaving their works, to a pressure of over 100 lbs. per square inch. In order to distinguish this hose from the common qualities usually sold, it is made of a chocolate colour, is stamped with the firm's trade mark, and has a guarantee label affixed to each roll sent out. Fine cloths being much stronger in proportion to weight than the heavy cloth made from inferior cotton, which is chiefly used on account of cheapness, Messrs. Charles Macintosh & Co. use a special fine cloth of great strength, and their *'one ply'* is made with two *'laps'*, *'two ply'* with three *'laps'* and so on.

In front of the stall we notice an India-rubber boat, suitable for fishing, for use as a pleasure boat, or for saving life. It is capable of carrying two persons, and yet, according to the statement affixed to it, will admit of being stowed away in an ordinary portmanteau.

Another noticeable exhibit is the pile of *Macintosh* tennis balls. The undoubted excellence of the best quality Melton-covered tennis balls produced by this firm, has for several years been acknowledged in many quarters In. future, each ball will bear the trade mark of the firm, and a strong effort is evidently being made to supply tennis players with the best ball

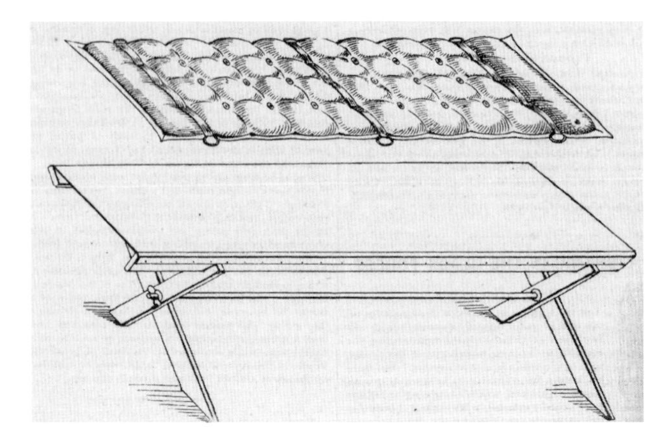

Fig. 15.2 The emergency or hospital air bed patented by Mr H. Waddington, manager at Charles Macintosh & Co. at that time.

in the market, under the name of the *Macintosh* tennis ball. The air bed[2] invented by Mr. H. Waddington has been so fully described in our columns, in September, 1885, and in February, 1887, that no further account is necessary [see Fig. 15.2]. We are glad to hear that it continues to meet with growing appreciation. We notice various fishing and sporting articles, and some waterproof rugs, rubber baths, &c., also some splendid specimens of fabrics prepared with the patent fast pile proofing for which this firm is so justly celebrated. Some excellent fast pile waterproof garments, both for ladies and gentlemen, are also shown.

Opposite to the varied and comprehensive collection we have just been looking at is another space also occupied by Messrs. Charles Macintosh & Co. It is backed by three large coloured pictures, representing scenes in the home of the rubber tree, on the banks of the Amazon-a settlement of rubber collectors and the native processes of procuring and smoking the rubber are graphically depicted. In front of these pictures a considerable space is protected by a low counter, and we see four girls at work, one making garments from waterproof fabric, one making tobacco pouches of sheet rubber, one engaged in making tennis balls, and another manufacturing elastic bands. Upon the counter are numerous interesting exhibits which we must barely enumerate. In a glass bottle is some of the rubber milk as it flows from the tree. Specimens of nuts used in smoking the rubber; specimens of Para and various kinds of rubber. These are followed by a number of glass jars containing a series of the products of the destructive distillation of rubber. Some specimens of well-made bandages, pouches, bands and tennis balls are also shown here. Some of the tobacco pouches are stamped with a view of the Manchester Exhibition building and grounds.

We must not omit to mention that between their two stalls Messrs. Charles Macintosh & Co., have laid down the vulcanized India-rubber floor covering known as '*Rabdotos*'. This material, specially recommended as '*dust arresting, mud preventing*', we have described in a previous number of the INDIA-RUBBER JOURNAL. It is warm, dry, and noiseless, and therefore very suitable for halls, passages, &c. It has a ribbed or fluted surface, and is of the same colour all through. It is supplied in long lengths, a yard wide, and, having a fluted surface, no joint is visible when lengths are laid side by side.

The following year they exhibited at the Glasgow International Exhibition and on finding the stall unattended a reviewer was drawn to comment:[3]

...the firm, having doubtless come to the conclusion that exhibitions were being overdone, merely contented themselves with placing their exhibit and leaving it to take care of itself'—a problem still facing many manufacturers today.

In 1889 the company took its products to Australia to exhibit at the Melbourne (Australia) Exhibition,[4] a far cry from its original shop in Charing Cross Road!

By the end of the 1880s there were at least 70 companies making mackintoshes of varying quality in Manchester alone. It is an old adage that one only gets what one pays for, so it is interesting to note that at this time one firm was producing mackintoshes with wholesale prices ranging from three shillings to over three pounds! The earlier designs had been made almost exclusively for men, or more properly gentlemen of a certain class, but now more colourful and stylish designs were becoming available for ladies, while the price range made some sort of protection available to manual labourers and office workers. Even so, three shillings with an additional retail mark-up was still a day's pay for many of the lower-paid workers.

The old problem of residual odour in fabrics which had been waterproofed using naphtha-based solutions of rubber broke out again in the early 1890s with one manufacturer, Joseph Mandleburg, registering as a trademark F.F.O., meaning Free From Odour. In 1893 the company tried to enforce by law its right to be the only company entitled to claim that their products were free from odour, but the application was rejected.

THE OLDEST HOUSE IN THE TRADE.

ORIGINAL INVENTORS

OF

Genuine "MACINTOSH" Waterproofs,

WHICH ARE

ODOURLESS and STAND ALL CLIMATES.

CAMBRIDGE STREET,
MANCHESTER, January 29th, 1892.

In consequence of recent advertisements and circulars cautioning the Trade against Infringements of certain Patents, dated 1888, we think it desirable to state that our **ODOURLESS** Garments are manufactured under patents and processes of earlier dates, which are altogether different from the ones referred to, and which have the advantage of perfectly Vulcanising the cloth in the first instance before the Garments are made up, rendering them quite **ODOURLESS** and unaffected by extremes of either heat or cold.

We can, therefore, recommend them as possessing the greatest advantages, being **ODOURLESS, DURABLE,** and of very **SUPERIOR FINISH.**

Every Garment bears our **Name and Trade Mark.**

FAC-SIMILE  OF LABEL.

Purchasers need not have the slightest fear that in buying them they are infringing the rights of any other person or firm, or that they are rendering themselves liable to any proceedings.

CHAS. MACINTOSH & CO., Ltd.

Fig. 15.3 Advert in self-defence by Charles Macintosh & Co. 1892. Note that the 'han(d)cock' trademark remains unaltered.

The response of the Macintosh Company was to take out a series of advertisements making its position clear (Fig. 15.3). The Macintosh patent (No. 4804 of 1883) reads, after explaining how the single or double texture is prepared:

I therefore expose this compound, fabric or substance in a stove room or other suitable place, to a temperature of 100 to 140° (F) for the purpose of maturing the manufacture and obtaining a further evaporation of the oil (*or naphtha*) which, while it remains, gives off a disagreeable effluvium and *so renders the garments free from smell.*

According to the label it sewed into all its mackintoshes and which is illustrated in its advertisement, it can be seen that Macintosh was using steam vulcanization and not the cold-cure process which had been purchased from Parkes in the 1840s. This would suggest that the fabric was cotton, although the company was selling mackintoshes made of such exotic materials as camel hair at that time, as both wool and silk, popular alternatives, were damaged by the steam process and were more usually cold-cured. The same argument applied to the more decorative ladies' mackintoshes, where bright organic dyes replaced the more sombre 'earth' colours.

The Macintosh company is not usually associated with cycle tyres but after the Dunlop Pneumatic Tyre Co. Ltd (soon to be renamed the Dunlop Rubber Company) was formed in 1889 to develop John Boyd Dunlop's pneumatic tyre patent—only to have the patent declared invalid a year later because of the earlier one of R. W. Thompson—there was a rush to be on board what could turn out to be a substantial industry. At a show in 1892 the following observation[5] was made about the Macintosh stand. The literary standard may be poor but the message is clear:

A firm whose connection with the rubber trade is from the beginning, and whose name has been so closely allied to garments that they bear the name, was, as may be expected, thoroughly represented. Starting with the Macintosh pneumatic, which has been considerably improved for 1893, in the unsightly rubber flaps, which previously secured it to the rims, being done away with, and the ridges of the outer cover being enclosed within the rim, thus giving a much neater appearance and a greater facility for repairs. The '*Holdfast*' tyre, which last season met with a very cordial reception among riders and manufacturers, has been lightened and further improvements have been effected; the '*Detachable*' tyre, which is attached to the rim by means of a flange; the '*Indestructible*' tyre, suitable for touring. The '*Macintosh*' Registered Cushion is now well known and needs no recommendation. These and numbers of other patterns, each ministering to the needs of special fancies, too numerous to mention, together with all kinds of handles, pedals, foot rests and other sundries, made up one of the most complete stands of the Rubber Section.

At a cycle show[6] in the early 1890s the same tyres were described as having many excellent qualities, but in 1894 came one patent battle too many for the company.[7] It was between The North British Rubber Co. with William Erskine Bartlett against Charles Macintosh & Co. and concerned the tyre which transformed bicycling: the 'Clincher'. This had a rim/tyre design which Bartlett had patented (and which remains the basis of today's design) in which the rim was cup-shaped in section rather than flat. The tyre, which had wire beading round its edges, fitted inside the shaped rim and was held in position by outward pressure from its own inflation. The system was not exactly as it is today but functioned by the same principle, and thus, for the first time, tyres could quickly be removed from the wheel and be repaired or replaced; obviously a tremendous advantage over earlier systems where the tyre was physically attached to the wheel and any repair was an involved procedure. The fight dragged on in court for a week in the Chancery Division of the High Court, after which time the judge, Mr Justice Romer, was completely confused and asked both parties if they would agree to an independent engineer's assessment of their respective systems. Four months later the engineer, Mr James Swinbourne, reported and the judge found against Chas. Macintosh. Although the argument went to appeal, the verdict was not changed and the company's last chance to break into the tyre business failed. With the internal combustion engine motor car just creeping over the horizon, this was a heavy loss to suffer.

Ironically, while the name of Thomas Hancock has been largely forgotten today, the main reason the name of Charles Goodyear remains known to the public is because of the Goodyear Tire and Rubber Company. This was founded in 1898, almost 40 years after Charles' death, by the Seiberling brothers in recognition of his work on vulcanization and as a marketing ploy. It has no actual connection with Charles or Nelson Goodyear.

By now there were many changes afoot in the rubber industry, not least the avalanche of new chemicals which could be added to a rubber mix before vulcanization to improve the product's properties or increase the speed of vulcanization. In 1904 it was discovered[8] that the addition of carbon black actually made a rubber vulcanizate stronger that it would otherwise have been, rather than reducing its strength as most inorganic materials added to bulk out the rubber did. It was soon discovered that different types conferred different properties on the finished product and another new industry was born: the manufacture of a wide range of carbon blacks. This period also saw the introduction of new mixing or compounding machinery, which had hardly changed from Thomas' and Walter's earlier designs. Workers soon discovered that adding significant levels of carbon black to rubber in a conventional masticator was a filthy business, and the search was on for an improved machine. In 1913 Werner Pfleiderer, a company which had been manufacturing machinery for mixing and kneading dough and had moved into manufacturing equipment for washing scrap rubber in the Malay peninsular,[9] launched the first closed internal mixer, and it was followed three years later by the eponymous 'Banbury', made by Fernley Banbury, which is still the most common mixer in use in the rubber industry today. Other advances since Thomas retired had been the introduction of steam-heated presses with shaped platens for forming and vulcanizing three-dimensional rubber products in one process. One could argue that this was merely an extension of Thomas' patent describing the use of heated moulds, although the temperature control available using super-heated steam made for vulcanization reproducibility and product consistency. Today electrical heating with full computer control of the whole mixing and vulcanizing cycle provides quality control undreamed of a century ago.[10]

It should also be remembered that in 1899 the first clean rubber arrived in the west from a plantation in Sri Lanka,[11] more than 50 years after Hancock had suggested that these should be set up. By The start of the First World War the output of 'plantation' rubber had exceeded that of the 'wild' material, and by 1920 over 90 per cent of rubber used worldwide was clean plantation rubber.[12] This must have been a great boon to the manufacturers, who were previously plagued by damage to mixer blades or rollers by grit or even stones. It could be mentioned here that it was not unusual for the Amazonian rubber collectors to 'build' their balls of raw smoked rubber (pelles) round a large stone to increase its weight and thus its price. However, the dealers soon cottoned on to this practice, and, as Fig. 15.4 shows, they used to cut the

Fig. 15.4 Checking rubber pelles for stones circa 1900.

pelles in half to check the core, although this would not eliminate grit or fine dirt.

The North British Rubber Company had grown from its humble beginnings in the shoe trade to be the largest rubber manufacturing company in the British Empire, and consolidations were taking place as fast as bankruptcies! In 1923 Charles Macintosh & Co. was taken over by the rapidly expanding Dunlop Rubber Company and the manufacture of rubber products continued on the site until February 2000, although the mill had to be rebuilt after the original one was destroyed in 1940 as a result of bombing raids.

The works were then earmarked for redevelopment as part of a private sector regeneration project for the area, known as the Southern Gateway to Manchester, the works are being converted for mixed residential and light commercial use and it is a tribute to Charles Macintosh that the area is named the Macintosh Village.

While Charles Macintosh & Co. were deep in the developing industrial heartlands of England, Stephen Moulton had chosen an unusual location for his factory, far removed from the more usual centres of industrial activity such as London, Birmingham, Manchester, Glasgow, and Edinburgh. It was a disused woollen mill, the Kingston Mill, at Bradford-on-Avon, Wiltshire. Nevertheless, it had a wealth of advantages, coal from Somerset, the river Avon alongside the mill to supply power and washing water, together with the closely adjacent Kennet and Avon Canal, and the Great West Road to provide access to London. It was also cheap, and contained within the 8-acre site was Kingston House, which would provide the family home.

In must be remembered that the foundations of the rubber-manufacturing industry were laid at a time when the Industrial Revolution was demanding ever-more powerful machinery and, the requirements of the rubber industry pushed those demands to, and sometimes beyond, the frontiers of technological development. Until 1821, when he installed a horse mill, Thomas Hancock's only source of power at Goswell Mews, London, was, literally, manpower and he could only process a few ounces of rubber at a time. In 1834 he had advanced to steam power at Manchester and later at Goswell Mews, while Charles Goodyear wrote in his Gum Elastic:

> It is want of adequate power and corresponding machinery for this purpose, and of that only, that the inventor is dissatisfied with the present state of the manufacture.

Fig. 15.5 The Iron Duke.

Neither Hancock nor Goodyear has much to say about the difficulties of making the required equipment, but when Moulton, in 1849, began equipping his factory, he installed what was probably the largest three-roll calender in the country: the Iron Duke (Fig. 15.5). The rolls were made at the Bilston foundry in Staffordshire and Bilston responded to Moulton's advice, which he was probably passing on from his American partners, the Rider brothers, by saying that he was extremely obliged for the hints concerning the grinding of the chilled rolls. He proposed that the rolls were to be placed on frames in the horizontal plane and turned round by cogwheels so that the surface of each rubs against the other. He thought that this would 'give an excellent and true surface without using emery'.[13]

Unfortunately the company did not have enough power available to turn the three rollers as the surfaces came ever-more into contact, and Bilston resorted initially to grinding just two of the rolls against one another. Two months later, on 3 April 1849, Moulton asked the obvious question:

How do you intend to grind the third? They must be ground together in the same frame, or otherwise, they will not be true.

After a further three months Bilston was moved to comment that after an expensive experience and a very serious loss ('if you could see the number of rolls strewing our foundry yard you would not complain') that they were compelled to abandon the machinery which they had been using and re-site the rollers in newer frames where they had more power.

Eventually the rolls were made and they were to the highest quality, but one does well to remember that Thomas' creation of a completely new industry required advances in technology that pushed the Industrial Revolution much further than it might otherwise have gone—and at a much greater speed!

After losing his patent battle with Charles Macintosh & Co. (pp. 142–144) Moulton was able to manufacture a wide range of goods but specialized in industrial and engineering applications.[14] It is noteworthy that to protect the interests of both Charles Macintosh and Co. and his nephew, Thomas ensured that the 'general licence' granted to Moulton excluded his manufacturing clothes and medical goods. Nevertheless, Moulton had a problem. As a small-scale manufacturer who could count his employees in a few tens, he could only compete with Macintosh and Co. on price. Quality, however good, was never enough to win orders against the competition and, even though there were no indications that Macintosh ran any sort of price-cutting war, this was never a satisfactory way to do business. The economic stimulus of the Crimean War saw him through the early years, and while his records show that he continued to produce rubberized fabrics, beds, and cushions through to 1880, it was in 1858 that the opportunity for a niche market presented itself. Moulton made contact with Messers George Spencer and Co., a London company that specialized in the design of components to meet the insatiable demands of the expanding rail networks, and a close informal partnership developed. The major products of Moulton's company were railway and carriage springs which, together with other railway-related products, grew from 30 per cent of output by value in 1860 to 85 per cent by 1890, the growth being due, in considerable measure, to Moulton's patented (1861) suspension unit, which consisted of a coiled spring embedded in a block of rubber.[15] Other areas of importance were hoses and sealing washers for valves. The company also flirted with rubberized conveyer belting in its early days, but had dropped out due to the intense competition and low profitability by the time of Moulton's death in 1880. Moulton kept his company small and efficient, ensuring employment for a small number of workers from the local village rather than risking expanding and doing battle with the likes of Macintosh and the rapidly growing North British Rubber Company. From the early 1850s through to the mid-1890s the number of employees stayed relatively constant between 50 and 80.[16]

In 1891 the company amalgamated with George Spencer of London to become George Spencer Moulton & Co. Ltd., and within a few years there was a degree of expansion as the company started producing motor tyres, something to which it devoted considerable effort during the Great War of 1914–18. By 1901 there were 164 employees. To put this figure into some perspective, in the 1861 census returns for England, Wales, and Scotland, those claiming to work in the rubber industry totalled fewer than 2,000 but by 1901 the figure exceeded 23,000. In this final year, over 40 per cent were female. The final decade of the nineteenth century saw a national growth in rubber workers of some 70 per cent, which could mostly be put at the feet of the bicycle and then motor tyre industries as well as the growing demand for electrical insulation By 1894, for instance, the new industry of bicycle tyre manufacturing accounted for more rubber than the footwear industry. In the Moulton factory the figure for labour expansion exceeded 100 per cent but, uniquely for this company, the majority of the figure was due to demands by the railway industry.[17]

With its emphasis on state-of-the-art design and manufacture it is not surprising that the Spencer–Moulton company also moved into supplying rubber components for aeroplane manufacturers. During the Second World War, Dr (later Sir) Alex Moulton, the great grandson of Stephen, was employed in the Engine Research Department of the Bristol Aeroplane Company. After the Second World War he was determined to move the family firm away from the manufacture of rubber suspension units for railway coaches and into automotive suspension. In 1956 the Spencer–Moulton Company became part of the Avon Rubber Company[18] and production ceased on the Bradford Mill site in 1993. However, undeterred, Alex Moulton started up a new company, Moulton Developments Limited, to design the suspension system for BMCs new small car, the Mini, the project of his friend, Sir Alec Issigonis. Alex already had experience in this field, as he had previously designed a rubber suspension unit for later models of the

Morris Minor, including a transverse-engine research model that never went into production. The Mini suspension further developed into the hydro-elastic and hydro-gas systems used initially in the 1100/1300 Austin/Morris series, the Allegro, and then the Rover 100 series followed by the MGF sports car. Alex Moulton also designed the eponymous Moulton Bicycle, which, like the Mini, used rubber suspension and unusually small wheels.

16

James Lyne Hancock & Co.

James Lyne Hancock & Co.—Thomas' original works at Goswell Mews taken over by J. L. Hancock as described earlier—development of the company—involvement with the Leyland Rubber Co.—death of James—firm left to his nephew, John Hancock Nunn—business continues to thrive under John's guidance. Thomas Hancock Nunn—his social pioneering and his involvement with Toynbee Hall—the flamboyant life of John Hancock Nunn—centenary of Thomas' founding of the company—reprint of the *Personal Narrative* but without the Appendix which contains the patent list—copy of the original (1857) book sent to King George V. The depression centred around 1830—problems of the company—contentious take over by the growing British Tyre & Rubber Co, (later B.T.R. Ltd)—death of John and loss of the fortune built up by Thomas and James Lyne Hancock.

As we have seen, Thomas Hancock passed his own company to his nephew, James Lyne Hancock (Fig. 16.1), son of Thomas' eldest brother James and his wife, Elizabeth née Lyne, in 1842. James had worked with his uncle for a number of years and was considered his most able assistant. It was obvious therefore that if Thomas intended to keep the business in the family it should pass to him. Another point in James' favour was that he held the same high principles and beliefs as Thomas and he was also a governor of the Wiltshire Society,[1] thereby confirming his belief in the power of education and apprenticeship as the main route out of poverty for deprived children.

Figure 16.2 shows the Goswell Mews site with the heavily outlined area being the Hancock works as it was in 1857, differentiated since the two blocks were leased from two separate landowners.[2] After the argument about hose and pipe manufacture with Charles Macintosh & Co., James had installed his own moulding and vulcanizing facilities at Goswell Mews, but space was at a premium. This plan was drawn up to

Fig. 16.1 James Lyne Hancock.

show the new premises which he intended to lease, consisting of all the units to the left of his current holding up to the edge of the plan.

For some time James had been friendly with James Quinn, and when the latter decided to establish a rubber manufacturing business in Leyland, near Birmingham, James Lyne agreed to second some of his workers to the East Street factory to train James Quinn's workers and to get the business under way.[3] With support like that it was not surprising that the factory flourished and in 1868 it moved into Leyland's old workhouse. James Quinn & Co. Ltd. described itself as being:[4]

> manufacturers of all kinds of india rubber articles, valves, sheets, buffers, washers, rings, cylinders, steam packing, hose tubing, india rubber machinery, belting, woven linen hose pipes for agricultural, fire brigade and mill purposes, and all india rubber articles used for engineering purposes, elastic steam rope, round or square, with core in the centre, and all kinds of water proof covers made to order, also water proof horse cloths etc.

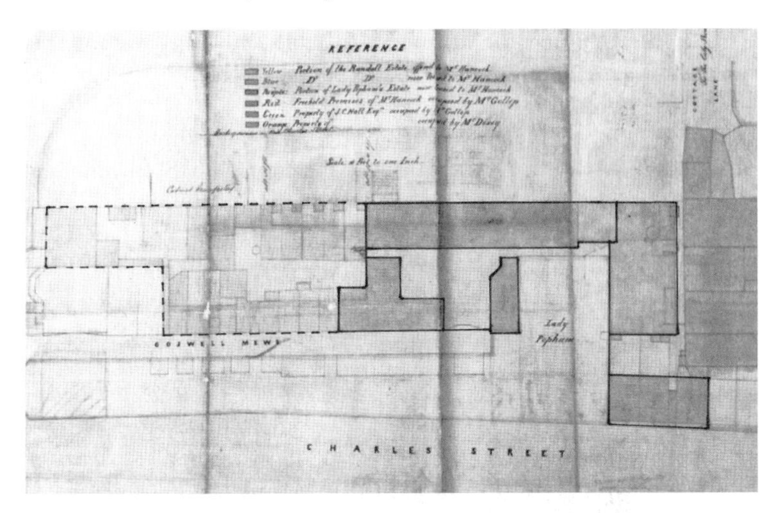

Fig. 16.2 Goswell Mews and surroundings in 1857.

Fig. 16.3 Advertisement for James Lyne Hancock—from circa 1900.

Looking at a contemporary advertisement by James Lyne (Fig. 16.3) it would seem that much of their two businesses were complementary rather than in direct competition, with Hancock tending towards the 'domestic' and Quinn, the 'industrial'. Hancock's detailed catalogue of 1867[5] confirmed this to be the case. Of 21 pages, only two could be considered to be of industrial applications—pipe seals, buffers or spring dampers, and mill engine drive bands—while one other was for brass fittings for garden hoses and one for ebonite sheets, tubing, taps, and photographic equipment.

James Quinn's company was renamed the Leyland Rubber Company after its founder died in 1883, and in 1898 it merged with the Birmingham Rubber Company to become the Leyland and Birmingham Rubber Company, which soon expanded its product range to include tyres, waterproof clothing, and surgical goods. Over the next half-century it became one of the great UK rubber companies and eventually, in 1969, merged with B.T.R. Ltd, although it continued to operate as a separate entity until it closed in the summer of 2002. Although it might be unrealistic to claim that it owed its success to the Hancock legacy and James Lyne Hancock, it is a fact that James Quinn was given a kick-start by James Lyne which could only have been to his advantage.

The Hancock Company continued to prosper under James Lyne, but he had no desire to move from Goswell Mews and, with space restricted, he was not interested in building an empire. His workers showed him the same appreciation that they had showed his uncle and that was sufficient. An amusing article[6] described one of the company's new products:

The position this firm occupies in the rubber trade is such that little praise is needed for any article they may bring out; but one of their recent productions is especially noticeable, as it promises to be a boon to users, and consequently should meet with a great sale by all rubber dealers. We refer to the Moulded Rubber Collar Stud. All men know the disadvantages of studs of the ordinary make. They are difficult to pass through the button-holes of the shirt and collar, they hurt the neck and fingers when inserted, they are uncomfortable, and from their unyielding nature give no freedom of play to the collar, and very soon wear out the button-holes, when the studs easily fall out, and are a constant source of worry; whereby, with Moulded Rubber Collar Studs, all these evils are overcome. The stud is made of solid rubber, moulded into the most approved form, which easily passes through the button-holes without hurting the fingers or neck. When in place the elastic nature of the rubber gives free play to the collar, and also prevents the wearing of the button-holes, and the peculiar adhesion of rubber also keeps it in place, and (the stud being very light) prevents it from falling out.

What man has not experienced the trouble of hunting for his collar stud when dressing, the only sure way of finding it being to put on one's thickest pair of boots, and before you have gone two steps you are certain to have completely crushed it beyond further service, but to grope about on the floor may take some considerable time, and exhaust a fair amount of patience before finding the straying stud. To all who have so suffered we would draw special attention to Messrs. Jas. Lyne Hancock's Moulded Rubber Collar Stud, which will be found indeed '*a boon and a blessing to men*.'

In the world of rubber manufacturing it would seem that that size is not everything!

James Lyne Hancock died in 1884, leaving an estate worth just over £300,000, a very considerable fortune by Victorian standards. Despite having two wives (in series, not in parallel) he left no children, but the inheritance was never in doubt. His cousin Sarah Hancock (Fig. 16.4) had married John Nunn (son of Thomas Nunn, one of Joseph Burrell's deacons). Their son, John Hancock Nunn, became the acolyte of James Lyne, playing the same role that James himself had played to Thomas Hancock so many years before. Inevitably, James named John Hancock Nunn as the principal legatee in his will and left him the company. He ran the business with energy and diligence until 1912, when it became a limited company.

John's day book,[7] covering the period from 1879 to 1920 survives and offers some interesting insights into the company. The vast range of products being manufactured by it during that period are grouped together under the heading 'general' but in June 1881 comes the first reference to 'tyres', with sales being just over one-tenth of turnover.

Fig. 16.4 Sarah Nunn, née Hancock.

By 1882 this figure had trebled in cash terms and risen to one-quarter of turnover. These would, of course, have been solid tyres. Although we saw in the previous chapter that the early 1890s saw the rise of the bicycle with its pneumatic tyres, an exhibition review of 1892[8] confirmed that the tyres being manufactured by John Nunn were solid ones. It seems unlikely to be a coincidence that with the advent of the 'clincher' tyre in 1894 the market for these solid tyres collapsed, dropping in John's day book to just a few pounds or even a few shillings per month, until production stopped in 1903. The company did take out a patent in 1904 for an 'elastic tire composed of a backing of hard rubber vulcanized to a more elastic tread', but it offered no competition to the pneumatic tyre and never went into production. The company briefly turned its attention to the manufacture of motor tyres during the First World War, although it never became more than a very small percentage of the business, probably on space grounds, and it was soon left to the specialist tyre companies to monopolize this business. The advert shown in Fig. 16.5 appeared in the press in the same month that sales of tyres were noted by John in his day book.

In 1885 a new category appeared in the day book and that was 'Gaussen'. It only generated a few pounds per month but continued through to the end of the book in 1920, so what is its significance? In

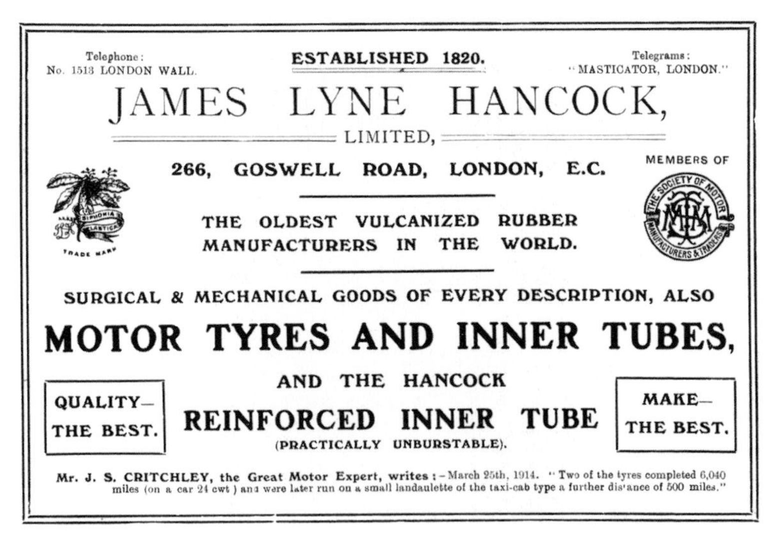

Fig. 16.5 Advertisement for James Lyne Hancock's tyres during 1914.

1881 David Gaussen had taken out a patent[9] for the manufacture of corrugated rubber sheeting as illustrated below (Fig. 16.6), and in the second paragraph he paid tribute to Thomas Hancock's patent of 1846 wherein Thomas described the use of moulds in which rubber could be vulcanized to take up and retain the shape of the mould. The illustration in Fig. 16.7 is also interesting, as it shows how corrugated tiles can be made to interlock using dovetail-type joints on as many sides as necessary. Although Gaussen lists every conceivable use he could think of for these corrugated sheets, the interest of John Hancock Nunn in this patent probably lay in the article referenced '*Fig. 10*' of Fig. 16.6, which is described as suitable for a mattress, bed, or seat, including in ambulances and operating theatres where they can be easily rolled up for storage and can be considered disposable if necessary. Such an article would fit nicely into the company's portfolio of medical and 'comfort' equipment such as those illustrated in Fig. 16.8. The reason for itemizing the sales was to calculate royalties which would be paid to Gaussen under the terms of the manufacturing licence which John Hancock Nunn must have taken out.

The Nunn family of John's generation continued to possess the character traits which had been passed down from James Hancock Senior through Thomas and James Lyne, but, however good and enlightened

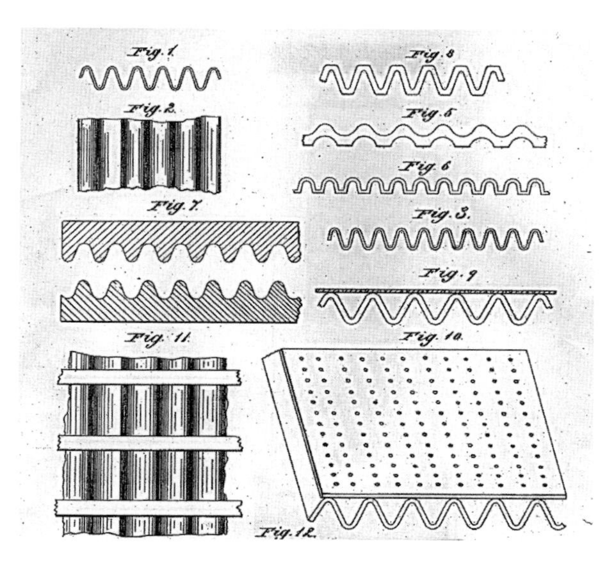

Fig. 16.6 Illustration from Gaussen's patent of 1881: various corrugations.

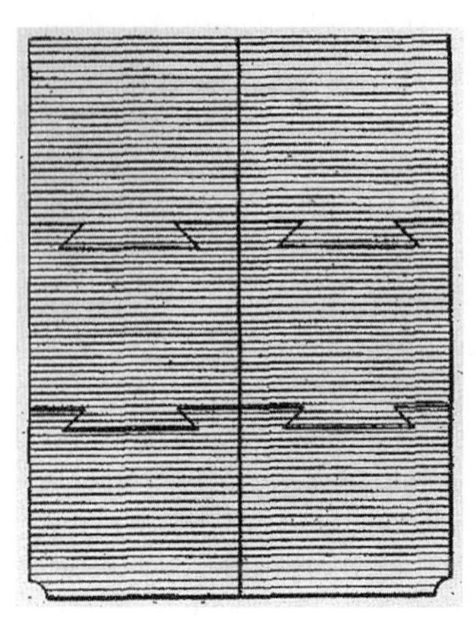

Fig. 16.7 Illustration from Gaussen's patent of 1881: 'dovetail' joints for corrugated floor tiles.

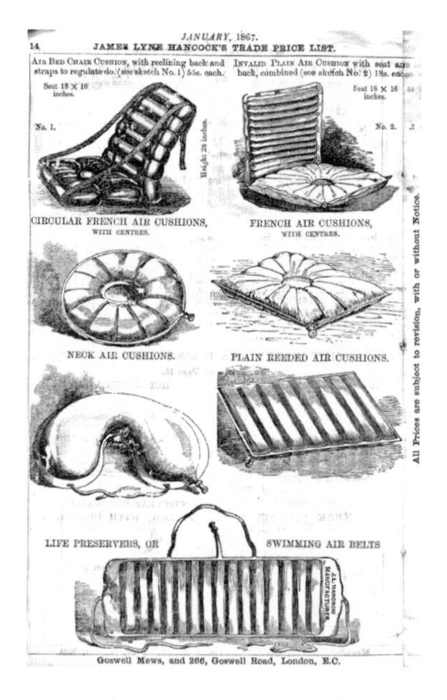

JANUARY, 1867.

14 JAMES LYNE HANCOCK'S TRADE PRICE LIST.

AIR BED CHAIR CUSHION, with reclining back and straps to regulate do. (see sketch No. 1) 55s. each.

INVALID PLAIN AIR CUSHION with seat and back, combined (see sketch No. 2) 18s. each.

Seat 18 × 16 inches.

No. 1.

Height 28 inches.

Seat 18 × 16 inches.

No. 2.

CIRCULAR FRENCH AIR CUSHIONS, WITH CENTRES.

FRENCH AIR CUSHIONS, WITH CENTRES.

NECK AIR CUSHIONS.

PLAIN REEDED AIR CUSHIONS.

LIFE PRESERVERS, OR SWIMMING AIR BELTS

J.L. HANCOCK MANUFACTURER

All Prices are subject to revision, with or without Notice.

Goswell Mews, and 266, Goswell Road, London, E.C.

Fig. 16.8 A page from the James Lyne Hancock catalogue of 1867.

an employer John was, he could stand firm when he had to. One such occasion arose in 1892 when an audit caught his head warehouseman, Thomas Bentley, stealing rubber from the company. John was not prepared to ignore the theft of a substantial amount of rubber, some £300 pounds worth, and the case progressed to the Quarter Sessions, where Bentley and his receiver were each sentenced to 16 months' hard labour.[10] In retrospect, for the period and for this value of goods, the punishment seems quite lenient!

On 1 January 1906 the following article appeared in the American journal *Rubber World*:

> The recent publication in the London papers of the name of Mr. T. Hancock Nunn, as a member of the Royal commission on the Poor Laws has led several people to assume that the gentleman named is identical with the Mr. J. Hancock Nunn, who is proprietor of the Vulcanized Rubber Works at 266 Goswell Road, London. This, however, I may say is not the case, though I am not prepared to say that there is no family connection. The trading name of the Goswell Road Rubber Works is James Lyne Hancock, the present proprietorship being as above mentioned in the hands of Mr. J. Hancock Nunn. The name will be familiar to readers of Dickens, as it was from a local habitation that Mr. Pickwick, on looking out of the window remarked that Goswell Street was on his right, Goswell Street was on his left, and the opposite side of Goswell Street was over the way. The business dates from 1820, its original founder being Thomas Hancock, of familiar memory in the trade. It can therefore claim to be the oldest established rubber business in the world, though only beating in this respect Messrs. Charles Macintosh & Co. by three years.

The writer omitted to note that Charles Dickens himself lived for a time in Goswell Street and, from his bay window, had a view of Thomas Hancock's works.

Thomas Hancock Nunn[11] (1859–1937) known to all as Tom Nunn, was actually the brother of John Hancock Nunn. He had no interest in the rubber business run by his brother but instead devoted his life to the poor. In the same year that James Lyne Hancock died leaving that colossal fortune, Toynbee Hall opened under Samuel Augustus Barnett, canon of St Jude's Church, and within a year Tom had found a base there. Situated in Commercial Street, Whitechapel, Toynbee Hall was Britain's first university settlement and the aim of its residents was both to educate the poor by holding university extension lectures and debates, and to give them the opportunity to continue their education past the school-leaving age. Many important institutions of social reform were launched from Toynbee Hall including the Workers

Educational Association in 1903 and later, the Citizens' Advice Bureau as well as the Child Poverty Action Group in 1965.

The autumn of 1888 was the time of Jack the Ripper, and Tom Nunn filled *The Times*'s letter columns with commentaries upon the appalling social conditions to be found in Whitechapel and how these contributed directly to criminality. He described local measures that he and his colleagues had helped put in place there to try to ameliorate the misery that such conditions engendered, wisely pointing out that the police alone could never defeat crime; only local communities organized by themselves for themselves would ever succeed—a lesson that many communities today might well heed. Tom Nunn's life however is another story all on its own, and here is not the place to tell it. However, it was beautifully summed up in the introduction to a memorial volume published on his death in 1937, and we cannot pass without repeating it.

> Thomas Hancock Nunn can best be described as a pioneer of social service, and like most pioneers his work was inspired by a sense of mission early conceived and persistent throughout life; the ideals which he formed were in their fundamentals very simple and unchanging, and were held with great strength of conviction . . . The first and central idea was both Christian and philosophical. As Christian it is embodied in the parable of the Good Samaritan, the lost pieces of silver, the Good Shepherd and the Prodigal Son. As philosophical it is expressed in Kant's doctrine of the infinite worth of the human soul. Human souls are liable to suffer from contact with a hard and evil world, liable to be stained with sin and corrupted by contact with evil, liable to be weakened, stunted and discouraged; but nothing in all this alters their infinite worth. The social servant's main and central task is to succour and support individuals. He may, and indeed must, be concerned also with many other things—housing, health, education, wages, poor relief etc etc, but he has to view these as affecting individuals, and of course must not sacrifice individuals to them. The State exists for the citizen, not the citizen for the State.

His life was a part of that great harvest which grew from the seeds planted by Huntington and nurtured by Burrell, protected and brought to fruition by the Hancocks and the Nunns.

While earlier members of the Hancock rubber dynasty had been relatively frugal with their cash, at least in terms of their own lives, John Hancock Nunn was determined to use his income and assets for his enjoyment.[12] In the first years of the twentieth century, he bought a splendid rambling old Sussex farmhouse, standing over a rabbit warren

Fig. 16.9 John Hancock Nunn with his wife, Rosa, and son Vivian at Lealands House.

of smugglers cellars. This piece of yeoman architecture dating back at least 300 years was razed to the ground and a new cement-rendered Italianate mansion was built in its place. It became known as Lealands House, Hellingly (Fig. 16.9), conveniently situated just off the old coach road from London to Brighton.

To this was added a sunken Italian garden and a large glass building (known locally as the Crystal Palace) in which summer dances could be held. A park was gradually assembled, a boating lake constructed (rubberized pitch lined of course), and neighbouring farms were added, until by the time he died in the same year as his brother Tom (1937) he owned, or so he claimed, all the land that could be seen from the roof of the water tower that crowned his mansion. It was an estate of several thousand acres. He lived the life of a wealthy country squire (Fig. 16.10).

John's great passion was driving four-in-hand coaches, a passion only open to the wealthy few, and he was a member of several syndicates

Fig. 16.10 A family gathering at Lealands House, August 1902. Significant characters are from left to right: gentleman in bowler, C. T. C. James, husband of Eliza; young boy kneeling, Rupert Hancock Nunn; behind Rupert is Harriet Hancock. The seated trio are Mrs Wynter-Blyth (John Hancock Nunn's mother-in law), John Hancock Nunn, and his wife, Rosa. Mr Wynter-Blyth is the bearded gentleman third from the right. The first lady on the left standing is Kate Hannah Nunn (Née Bourne), wife of Tom Nunn.

which ran those exotic carriages. The purpose of their passion was gastronomic as well as nostalgic, for the passengers were wont to stop for very long periods at the best hotels along the route to refresh themselves and the luncheon menu was quite as important as the itinerary. John took great pleasure in carrying his close friends to Ascot and Epsom racecourses on race days on one of his coaches. While his return to the enjoyment of horse-drawn carriages at the time of the birth of motorized vehicular transport had a certain irony in view of the work of his great uncle Walter, it is also amusing that he was wont to trouble the editor of *The Times* with irritable letters about such menaces as motorists 'scorching' along country roads, no doubt on those new-fangled pneumatic rubber tyres!

In 1920, with the company still known as James Lyne Hancock Limited, the firm reprinted Thomas' *Personal Narrative* to celebrate

PERSONAL NARRATIVE

OF

THE ORIGIN AND PROGRESS

OF THE

CAOUTCHOUC

OR

INDIA-RUBBER MANUFACTURE

IN

ENGLAND.

BY

THOMAS HANCOCK.

Published by
Longman, Brown, Green, Longmans & Roberts,
London.

—

1857.

REPRINTED
BY ORDER OF THE DIRECTORS OF
JAMES LYNE HANCOCK, LIMITED.
VULCANIZED RUBBER WORKS,
266, GOSWELL ROAD,
LONDON, E.C.

—

1920.

N.B. **The appendixes containing the specifications of Mr Hancock's patents have been omitted in this edition.**

Fig 16.11 Title page of James Lyne Hancock's reprinted edition of Thomas Hancock's *Personal Narrative.*

100 years of the company (Fig. 16.11). This edition did not include the specifications of Thomas' 14 patents or the mass of statistical data which had been present in the original publication but which Thomas had also published privately in 1853 as a separate book. They did, however, include three composite illustrations of life within the factory at that time (see Figs 16.15–16.17). The company also treated its workers to a day out at the seaside (Fig. 16.12)!

Although Thomas sent copies of his *Narrative* to many people, there is no evidence that Queen Victoria or, perhaps more appropriately, Prince Albert received one. With the printing of the centenary edition the chairman of the company, Sir Charles Inigo Thomas GCB, obviously felt a need to rectify this omission and the letter[13] illustrated in Fig. 16.13 shows that the opportunity was taken to present King George V with a copy not of the centenary edition but with one of the original 1857 *Narrative* print runs. The signature is that of Lord Stamfordham, Private Secretary to the King.

THE START. ON JUNE 12TH 1920.
From Works of JAMES. LYNE. HANCOCK. LTD. For Brighton.
To Celebrate Centenary of Rubber Manufacture in England.

Fig. 16.12 A day out for the workers!

BUCKINGHAM PALACE

11th. June, 1920.

Dear Sir Inigo Thomas,

I have received and laid before the King the original copy of Thomas Hancock's "Personal Narrative of the Origin and Progress of India Rubber Manufacture in England", which His Majesty is interested to read, and for which I am commanded to express his best thanks.

Yours very truly,

Stamfordham

Sir C. Inigo Thomas,
G.C.B.,
Chairman,
Messrs. James Lyne Hancock,
Limited,
266 Goswell Road,
E.C.1.

Fig. 16.13 Letter from Buckingham Palace to James Lyne Hancock & Co., 1920.

From the late 1920s to the mid-1930s the Great Depression began to take its toll, although the company did manage to survive this period intact. The family still owned a large shareholding, but other events had distracted their attention from the milch cow which sustained them. John Hancock Nunn was growing old and, for the first time, the family had to contemplate death duties, which had been introduced at the end of the nineteenth century, luckily for them after the death of James Lyne Hancock. They also suffered a reduced income from their farms with the onset of an agricultural depression. John Hancock Nunn's two sons had found careers in fields other than rubber: the elder son, Rupert, after serving with distinction in the Great War had become a director and partner in Constable and Maud, a leading London firm of estate agents, while Vivian had trained as a barrister.

While the Hancock Nunn family continued living as country squires amid a now crumbling agricultural empire, there were those with very

Fig 16.14 The Hancock Nunn factory, 1920. Views of mixing mills and calenders. Spreading shops and hose-making department.

sharp eyes who were paying close attention to the worsening political landscape in Europe. They were inclined to believe that there was to be another world war, and in this war those who controlled strategic materials would be well placed to make a substantial fortune. Across the industrial landscape speculators were quietly assembling large complexes and among these was Sir Walrond Sinclair, whose specialty was rubber companies which he collected through his conglomerate, the British Tyre and Rubber Company Ltd., so named in 1934 when it was floated as a UK public company, although, as we have seen, certain elements of it date back to rubber-related businesses founded in the nineteenth century. In 1955 it became B.T.R. Limited.

Somehow Sir Walrond Sinclair managed to gain control of James Lyne Hancock Ltd. Exactly how he was able to do this and gain control of so many other rubber companies is open to question—a question that the Hancock Nunns were totally unable to answer because they could not prove their suspicions. All it would be safe to say here is that they instigated a protracted correspondence with B.T.R. Ltd. in a determined attempt to elicit certain information concerning financial

Fig 16.15 The Hancock Nunn factory, 1920. Views of vulcanizing and moulding shops, stores, washer, and ring-cutting departments.

statements in the company's accounts which they believed to be fraudulent, but in this they ultimately failed.[14] A series of files was submitted to the Director of Public Prosecutions asking him to intervene and a Board of Trade Enquiry was mooted, but in the end nothing came of these flailings. Sir Walrond Sinclair KBE went on to become chairman of a number of companies, including, by a strange irony, the India Rubber, Gutta Percha and Telegraph Works, which incorporated Charles Hancock's old company. By acquiring these concentrated rubber interests, he became Rubber Controller at the Ministry of Supply in 1941, a very wealthy man, and, of course, a pillar of the Establishment.

With the death of John Hancock Nunn in 1937 the fortunes of the Hancock Nunn family spiralled downhill. First death duties and later the agricultural depression slowly but persistently nibbled away at the estate. Constable and Maud went bankrupt so Rupert Nunn lost everything. Vivian managed to regain Lealands House after the war, from the War Ministry, which had sequestered it, but to remain there after the war cost him most of the remaining lands. The house was finally

Fig 16.16 The Hancock Nunn factory, 1920. Make-up area for small items: rings, washers, cones, football bladders, and water bottles, etc.

sold in the 1980s, when Vivian's widow, Eileen, went into a nursing home. What was left of the great rubber fortune, now shrunk almost to nothing, was put into trust to contribute to the upkeep of their only son, who was in sheltered accommodation.

And so the story of the rise and fall of the Hancock fortune draws to a close. There is only one thing left to mention. One of the daughters of John Hancock has only received a passing mention in this story: Eliza (Fig. 16.17). She was married in 1857 to Charles Clement James, great grandfather of Francis James. They had one child, Charles Thomas Clement James.

He was brought up by his mother in the expectation that he would inherit the rubber business from James Lyne Hancock, but was to be disappointed. He wrote 27 novels, was a double bigamist, and spent a year in Wormwood Scrubs, but that too is another story for another time.

James Lyne Hancock Ltd produced independent accounts until the year 1960, when it was finally and completely absorbed into the parent company. In 1999 B.T.R. Ltd. merged with Siebe to form B.T.R. Siebe plc, which was later renamed Invensys plc.

Fig 16.17 Eliza James, née Hancock.

Epilogue: Thoughts on a Dynasty

1749 saw the first performance of the Royal Fireworks Music, composed by Fredric Handel and accompanied by what was intended to be a magnificent pyrotechnic display orchestrated by the Italian virtuoso, Giovanni Servandoni. A shower damped down the proceedings and only a few of the fireworks performed as planned.

Just over 50 years later William Congreve designed and built the Congreve rocket with its new propellant and various versions of these were used extensively against the French armies of Napoleon as well as against the United States in the war of 1812. There can be little doubt that the Hancock family would have been familiar with illustrations relating to their use in the Napoleonic war and to the burning of Copenhagen and Washington DC. Indeed, a line in *The Star Spangled Banner*, written in 1814 by Francis Scott Key reads 'And the rockets' red glare, the bombs bursting in air'. They would also have known that the rockets were unpredictable in both the direction they travelled and the damage they did when they arrived. Like Servandoni's fireworks they could let you down.

The story of the Hancock family seems to compare well with the appearance of a pyrotechnic rocket in the dark skies of post-Napoleonic England. The touchpaper was lit at the Marlborough Academy, the explosive energy that lifted it to the skies arose from a potent mixture of native genius and a living Christian faith. The showers of brilliance that illuminated the night skies, and fell (mostly) like blessings on the human race, were the results of continuing diligent experiments, and then came the end of all rockets, the silent falling of the stick and the burnt-out shell with its consignment to the litter bin of history.

Thomas was the pyrotechnic that performed outstandingly in all aspects with an ongoing sequence of flashes of genius, propelled even higher and brighter by the explosive inputs from James Lyne and John Hancock Nunn. William, Walter and Charles all promised so much with their initial pyrotechnic displays but after brief explosive flashes they were damped down by their different character deficiencies and faded away. Other members of the family added to the background sparkle in the skies with their diverse and fascinating characters but in the end the display was over and history has since ignored one of the greatest families of genius produced during the Industrial Revolution.

Hopefully this book has corrected this.

References and Source Documents

Throughout the book there are many quotes attributed to Thomas and Walter Hancock. Rather than distract the reader with numerous individual references it can be assumed that they came from their respective narratives unless otherwise indicated. The full titles of their narratives are

Hancock, T. *Personal Narrative of the Origin and Progress of the Caoutchouc and India-rubber Manufacture of England.* Longman, Brown, Green, Longmans & Roberts, London 1857.

Hancock, W. *Narrative of Twelve Years Experiments Demonstrative of the Practicability and Advantages of Employing Steam carriages on Common Roads.* John Weal, High Holborn & J Man, Cornwall, 1838.

Thomas' *Narrative* was reprinted in 1920 by James Lyne Hancock Ltd. but contains only Thomas' text and his section on Mechanical Applications of Vulcanized India-rubber. It does not contain a mass of information on rubber production and imports/exports worldwide nor does it contain the list of patents, the headings of which are shown here in Appendix I. For the full texts of these one needs the 1857 *Narrative* or

Hancock, T. *Specifications of Fourteen Patents Granted by Geo. IV. Will. IV. and Victoria from April 29, 1820 to Dec. 30, 1847.* George Barclay, Leicester Square (for private circulation) 1853.

The text of Thomas' book is also available on-line at http://www.lakelandelements.com/rainwearhistory/hancock_personal_narrative.htm/.

On occasions when it is necessary to reference the text for other than direct quotations the two narratives will be referenced simply as the appropriate *Narrative.*

Notes

CHAPTER 1

1. Wordsworth, C. Old Marlborough Pamphlets, Herbert G Perkins Times Printing Office, Marlborough, 1904.
2. Ibid.
3. Hancock archives.
4. Cunnington, Presentments of the Grand Jury of the Quarter Sessions, Marlborough 1706–51, Devizes, 1929.
5. Hancock archives.
6. *Cambrian*, 24 February 1804.
7. *The Times*, 8 July 1808.
8. Hancock archives.
9. Wordsworth, C. letter to the *Marlborough Times*, May 1903.
10. Gibbs, R. *History of Aylesbury, Bucks Advertiser and Aylesbury News*, Aylesbury, 1889.
11. Ibid.
12. Ibid.
13. http://en.wikipedia.org/wiki/Year_Without_a_Summer.
14. Gibbs, R. *History of Aylesbury*, Robert Gibbs, Bucks Advertiser and Aylesbury News office, Aylesbury, 1889.
15. Viardot, L. *A Brief History of the Painters of All Schools*, Read Books, London, 2007.
16. *Salisbury Gazette*, 23 April 1818.

CHAPTER 2

1. Benson, W.A. *Discourse Occasioned by the Death of the Late Thomas Hancock*, Printed privately, July 1865. Hancock archives.
2. Brant, C. William Huntington. In: *The Online Oxford Dictionary of National Biography*, ed. Lawrence Goldman.
3. Philpot, J.C. *William Huntington,* http://www.gracegems.org/18/p-Huntington.htm.
4. Ibid.
5. Post Office Directory, 1816.
6. Robson's Directory, 1820.
7. Anon. Exhibition of Napoleon's carriage in London, http://www.georgianindex.net.
8. Fuller, S & J. Promotional inscription related to Fig. 2.1. 1920.

9. Loadman, J. *Tears of the Tree.* Oxford University Press, Oxford, 2005.
10. De Chasseloup Laubat, F. *François Fresneau, Seigneur de la Gataudière,* Les Petits-Fils de Plon et Nourrit, Paris 1942.
11. Ibid.
12. Priestly, J. *A Familiar Introduction to the Theory and Practice of Perspective,* London. 1770, quoted by Dawson, T.R. *History of the Rubber Industry,* eds. P. Schidrowitz and T. R. Dawson, W Heffer & Sons, 1952.
13. Dawson, T.R. *History of the Rubber Industry,* eds. P. Schidrowitz and T. R. Dawson, W Heffer & Sons, Cambridge,1952.
14. Hancock archive.
15. Hancock, T. *Narrative.*
16. Hancock archive.
17. English patent 5208 (1825).

CHAPTER 3

1. Norwich Society of Artists, Catalogue, 1818–21.
2. Hancock, T. *Narrative.*
3. Hancock, W. *Narrative*, 1838.
4. *History of the Automobile,* http://www.sapiensman.com/old_cars.
5. James, F. *Walter Hancock and His Common Road Steam Carriages.* Laurence Oxley, Arlesford, 1975.
6. *Glasgow Herald*, 1 August 1834.
7. Emmerson, G. *John Scott Russell.* Murray, London, 1977.

CHAPTER 4

1. *Reading Mercury*, 31 July 1826.
2. Ibid, 20 February 1926.
3. Hancock archive.
4. British Library Print Collection.
5. Suffolk Record Office.
6. Hancock archive.
7. Ibid.
8. Royal Academy Exhibition Catalogue, 1824.
9. Hancock archive.
10. Ibid.
11. Macintosh, C. *Biographical memoir of the Late Charles Macintosh F.R.S. of Campsie and Dunchattan,* Printed for private circulation by W.G. Blackie & Co., Villafield, 1847.
12. Love and Barton, *Manchester as it is*, Love and Barton, Manchester, 1839.
13. Ibid.
14. Hancock, T. *Narrative.*

15. Woodcock, E. Letter to Thomas Hancock dated 1857. Hancock archive.
16. Hancock, T. Dough Waterproofing; patent enrolled 17 October 1837.
17. Plaque in Ludgvan Church, near Marazion, Cornwall.
18. Correspondence, Hancock archive.
19. James, J.L.B. *Memoir,* unpublished, Hancock archive. See Chapter 9.
20. Hancock archive.

CHAPTER 5

1. Love & Barton. *Manchester as it is.* Love and Barton, Manchester 1839.
2. British History Online, taken from *A History of the County of Middlesex*, vol. 8., eds. T. F. T Baker and C. R. Elrington, 1985.
3. Tythe map 1848 and 1841/51 census returns, London Borough of Hackney Archives.
4. Policy No. MS 11936/551/1238435, Guildhall Library.
5. James, J.L.B. *Memoir,* unpublished, Hancock archive. See Chapter 9.
6. Correspondence between Mr Bettely and Thomas Hancock, Hancock archive.
7. Hancock, W.C. in typescript of lecture of 3 March 1924 and Chairman's response, unpublished, Hancock archive.
8. Alexander, J.H. *More than Notion.* Fauconberg Press, London, 1967.
9. Idem.
10. Hancock archive.
11. Macintosh, C. *Biographical Memoir of the Late Charles Macintosh F.R.S. of Campsie and Dunchattan,* Printed for private circulation by W.G. Blackie & Co., Villafield, 1847.
12. Loadman, J. *Tears of the Tree.* Oxford University Press, Oxford, 2005.
13. Hancock, T. *Narrative.*
14. Hancock, W. *Narrative.*
15. *Bury and Norwich Post*, 9 October 1839.
16. *Cambridge Chronicle*, 8 October 1839.
17. Hancock, W. *Narrative.*
18. English Patents 7037 (21 March 1836) and 8765 (14 January 1841).
19. *Bewley* v *Hancock*, Chancery Proceedings 1855–60.

CHAPTER 6

1. Hancock, T. *Narrative.*
2. Slack, C. *Nobel Obsession*. Hyperion, New York, 2002.
3. Goodyear, C. *Gum Elastic. India Rubber Journal*'s facsimile reproduction, Maclarans and Sons, London 1937.
4. Ludersdorff, F.W. *The Solution and Manufacture of Elastic resin called Gum Elastic for the Manufacture of Air- and Water-tight Materials.* J. W. Bioke, Berlin, 1932.

5. Lunn, R.W. *History of the Rubber Industry,* eds. P. Schidrowitz & T. R. Dawson, W Heffer & Sons, Cambridge, 1952.
6. Peirce, Revd B.K. *Trials of an Inventor; Life and Discoveries of Charles Goodyear.* Carlton & Porter, New York, 1866.
7. Loadman, J. *Tears of the Tree.* Oxford University Press, Oxford, 2005.
8. *1848–1948 A Hundred Years of Rubber Manufacture.* Privately published by George Spencer Moulton & Co. Bradford-on-Avon, 1948.
9. Hancock, T. *Narrative.*
10. *1848–1948 A Hundred Years of Rubber Manufacture.* Privately published by George Spencer Moulton & Co. Bradford-on-Avon, 1948.
11. Slack, C. *Nobel Obsession.* Hyperion, New York, 2002.

CHAPTER 7

1. Hancock, T. *Narrative.*
2. Hancock archive.
3. Schidrowitz, P. *Prevulcanization of Latex*, British Patent 193451, 1921.
4. http://www.plastiquarian.com/parkesine.htm (website of the Plastics Historical Society).
5. Hancock, T. *Narrative.*

CHAPTER 8

1. Beauchamp, K.G. *History of Telegraphy.* IET, London, 2001.
2. Black, R.M. *History of Electric Wires and Cables.* IET, London, 1983.
3. Eaton. B.J. *Wild and Plantation Rubber Plants—Gutta Percha and Balata in History of the Rubber Industry,* eds. P. Schidrowitz & T. R. Dawson, W Heffer & Sons, Cambridge, 1952.
4. Glover, W. *British Submarine Cable Manufacturing Companies* at http://www.atlantic-cable.com/CableCos/BritishMfrs/.
5. *Bewley* v *Hancock*, Chancery Proceedings, 1855–60.
6. *Art Union Journal*, December, 1843.
7. Hancock, T. *Narrative.*
8. *Bewley* v *Hancock*, Chancery Proceedings, 1855–60.
9. Ibid.
10. Ibid.
11. Ibid.
12. Ibid.
13. Ibid.
14. Ibid.
15. Hancock archive.
16. Letter ref. reel 48/1/41/155, Cornwall Record Office, Papers of Harvey & Co. of Hayle. Adam Matthews Publications, Marlborough.

17. *Chelmsford Chronicle*, 21 May 1852.
18. Hancock archive.
19. *Chelmsford Chronicle*, 21 May 1852.
20. *Chelmsford Chronicle*, 28 May, 1852.
21. Glover, W. *British Submarine Cable Manufacturing Companies* http://www.atlantic-cable.com/CableCos/BritishMfrs/.
22. Siemens corporate archives athttp://w4.siemens.de/archiv/en/dokumente/werner_von_siemens_en.pdf.
23. Lawford, G.L. and Nicholson, L.R. *The Telcon Story 1850–1950.* The Telegraph Construction and Maintenance Co., London, 1950.

CHAPTER 9

1. This chapter is based on a collection of correspondence between Thomas Hancock, Charles Macintosh & Co. and James Lyne Hancock, copies of which survive in the Hancock archive.

CHAPTER 10

1. Korman, R. *The Goodyear Story: An Inventor's Obsession and the Struggle for a Rubber Monopoly. Encounter Books*, San Francisco, 2003.
2. Graves, A. *The Royal Academy of Arts—A Complete Dictionary of Contributors and their Works from its Foundation in 1769 to 1904.* Henry Graves & Co., London 1906.
3. Korman, R. *The Goodyear Story: An Inventor's Obsession and the Struggle for a Rubber Monopoly.* Encounter Books, 2003.
4. Goodyear, C. *Gum Elastic.* Maclarans and Sons, London 1937. India Rubber journal's facsimile reproduction.
5. Miscellaneous Correspondence Collection, Royal Botanic Gardens, Kew.
6. Hancock archive.
7. Museum Entry Book, Royal Botanic Gardens, Kew.
8. Loadman, J. *Treasures of Kew* in Kew Magazine, Royal Botanic Gardens, Kew, London, 2005.

CHAPTER 11

1. *1848–1948 A Hundred Years of Rubber Manufacture,* Privately published by George Spencer Moulton & Co. Bradford-on-Avon, 1948.
2. *Mechanics Magazine*, 29 May 1852.
3. Ibid, 12 June 1852.
4. Hancock, T. *Narrative.*
5. *1848–1948 A Hundred Years of Rubber Manufacture,* Privately published by George Spencer Moulton & Co. Bradford-on-Avon, 1948.

6. Woodruff, W. *The Rise of the British Rubber Industry During the Nineteenth Century.* Liverpool University Press, Liverpool, 1958.
7. Selected papers from the *Kew Bulletin—III Rubber,* Royal Botanic Gardens, Kew, HMSO, London, 1906.

CHAPTER 12

1. Letters referred to in this chapter are in the Hancock archives.
2. Hancock archives.
3. Hancock archives.

CHAPTER 13

1. Benson, W.A. *Discourse Occasioned by the Death of the Late Thomas Hancock*, printed privately, July 1865. Hancock archives.
2. Quoted by Benson, W. *A Discourse Occasioned by the Death of the Late Thomas Hancock,* Printed privately, July 1865. Hancock archives.
3. Hancock archive.
4. Prosser, R.B. and McConnell, A. Thomas Hancock in *The Oxford Dictionary of National Biography Online.* Oxford University Press, Oxford, 2007.
5. Hancock, T. Last Will and Testament, Principle Registry of the Probate Divorce and Admiralty Division of the High Court of Justice. Proved with three Codicils 2 June 1865. Hancock archives.
6. Hancock archive.
7. Hancock archive.
8. Hancock archive.
9. Porritt, B.D. and Rogers, H. 'The Life and Work of Thomas Hancock', Lecture given to The Institution of the Rubber Industry, 3 March 1924. Hancock archives.
10. Hancock, W. Chairman's remarks given after the lecture 'The Life and Work of Thomas Hancock', referenced above, Hancock archive.

CHAPTER 14

1. James, J.L.B. *The Great Aunts*, unpublished, 1950, Hancock archives.
2. http://www.plastiquarian.com/plaque04.htm (website of the Plastics Historical Society).

CHAPTER 15

1. *The India Rubber and Gutta Percha and Electrical Trades Journal*, 8 June 1887.

2. Ibid, 8 February 1887.
3. Ibid, 8 September 1888.
4. Ibid, 8 January 1890.
5. Ibid, 9 January, 1893.
6. Ibid, 8 December 1891.
7. Ibid, 8 March 1894, 9 July 1894.
8. Loadman, J. *Tears of the Tree*, Oxford University Press, Oxford, 2005.
9. Ackers, C. E. *The Rubber Industry in Brazil and the Orient*, Methuin & Co. Ltd., London, 1914.
10. Patel, S.R. and Halladay, J.R. *Computer Control in Rubber Processing.* Rubber World, November 1994.
11. Browne, E.A. *Peeps at Industries—Rubber.* Adam & Charles Black, London 1912.
12. Loadman, J. *Tears of the Tree*, Oxford University Press, Oxford. 2005.
13. Woodruff, W. *The Rise of the British Rubber Industry During the Nineteenth Century.* Liverpool University Press, Liverpool, 1958.
14. *1848–1948 A Hundred Years of Rubber Manufacture*, Privately published by George Spencer Moulton & Co. Bradford-on-Avon, 1948.
15. Ibid.
16. Woodruff, W. *The Rise of the British Rubber Industry During the Nineteenth Century*, Liverpool University Press, Liverpool, 1958.
17. Ibid.
18. *Avon Rubber plc. A Period of Rapid Growth—Part 1.* http://www.avon-rubber.com.

CHAPTER 16

1. Apprenticeship Records, The Wiltshire Society. Appendix B—Governors of the Wiltshire Society, *1817–1921*, Wiltshire & Swindon Archives.
2. Hancock archive.
3. Online records of The Leyland Historical Society, http://www.houghton59.fsnet.co.uk/lanes%2009.htm.
4. Ibid.
5. Han cock archive.
6. *The India Rubber and Gutta Percha and Electrical Trades Journal*, 8 April 1890.
7. Hancock archive.
8. *The India Rubber and Gutta Percha and Electrical Trades Journal*, 8 December 1892.
9. USP 239159, 22 March 1891.
10. *The India Rubber and Gutta Percha and Electrical Trades Journal*, 9 May 1892.

11. Anon. *Thomas Hancock Nunn—The Life and Work of a Social Reformer.* Baines and Scarsbrook Ltd, London, 1942.
12. Hancock archive.
13. Ibid.
14. Ibid.

APPENDIX I

The 14 patents of Thomas Hancock 'for the treatment and application of INDIA RUBBER'

Although the 14 patents taken out by Thomas Hancock appear in his relatively rare 1857 *Narrative,* they were omitted from the (relatively) more common 1920 reprint. They also appear in the extremely rare book privately published by Thomas in 1853 which is dedicated to them. For those who have an interest in studying them in more detail, their titles, introductions and dates are listed so that they may be identified by the UK Intellectual Property Office.

The Fourteen patents of Thomas Hancock 'for the treatment and application of INDIA RUBBER'

ARTICLES OF DRESS. Specification of Patent granted to THOMAS HANCOCK, Stoke Newington, Middlesex, Esquire, for an Improvement in the Application of a certain Material to various Articles of Dress and other Articles, that the same may be rendered more Elastic.

Dated April 29, 1820. Enrolled August 8, 1820

PITCH AND TAR. Specification of Patent granted to THOMAS HANCOCK, Stoke Newington, Middlesex, Esquire, for an Improvement in the Preparation for various useful purposes of Pitch and of Tar, separately or in union, by an Admixture of other Ingredients with either or both of them.

Dated March 22, 1823. Enrolled September 17, 1823

LEATHER BY LIQUID. Specification of Patent granted to THOMAS HANCOCK, Stoke Newington, Middlesex, Esquire, for Improvements in the Method of Making or Manufacturing an Article which may be, in many instances, substituted for Leather, and be applied to various other useful purposes.

Dated Nov. 29, 1824. Enrolled May 28, 1825

LEATHER BY SOLUTION. Specification of Patent granted to THOMAS HANCOCK, Stoke Newington, Middlesex, Esquire, for a New or Improved Manufacture, which may, in many instances, be used as a Substitute for Leather, and otherwise.

Dated March 15, 1825. Enrolled September 14, 1825

ROPES AND CORDAGE. Specification of Patent granted to THOMAS HANCOCK, Stoke Newington, Middlesex, Esquire, for an Improvement or

Improvements in the Preparation or in the Process of Making or Manufacturing of Ropes or Cordage, and other Articles from Hemp, Flax, and other Fibrous Substances.

Dated March 15, 1825. Enrolled September 14, 1825

ORNAMENTS, ETC. BY LIQUID. Specification of Patent granted to THOMAS HANCOCK, Stoke Newington, Middlesex, Esquire, for Improvements in the Manufacture of certain Articles of Dress or Wearing Apparel, Fancy Ornaments, and Figures, and in the Method of rendering certain Manufactures and Substances in a degree, or entirely, impervious to Air and Water, and of protecting certain Manufactures and Substances from being injured by Air, Water, or Moisture.

Dated August 5, 1830. Enrolled October 5, 1830

EXPANDING CUSHIONS. Specification of Patent granted to THOMAS HANCOCK, Stoke Newington, Middlesex, Esquire, for an Improvement or Improvements in Air-Beds, Cushions, and other Articles Manufactured from Caoutchouc or Indian Rubber, or of Cloth or other flexible material, coated or lined with caoutchouc or Indian Rubber.

Dated June 4, 1835. Enrolled December 4, 1835

DOUGH WATERPROOFING. Specification of Patent granted to THOMAS HANCOCK, Stoke Newington, Middlesex, Esquire, for an Improvement or Improvements in the process of rendering Cloth and other Fabrics partially or entirely impervious to Air and Water by means of Caoutchouc or India Rubber.

Dated April 18, 1837. Enrolled October 14, 1837

DOUGH SHEETS. Specification of the Patent granted to THOMAS HANCOCK, of Goswell Mews, in the County of Middlesex, Patent Waterproof Cloth Manufacturer, for Improvements in the Method of Manufacturing or Preparing Caoutchouc, either alone or in Combination with other Substances.

Sealed January 23, 1838. Enrolled July 28, 1838

VULCANISING. Specification of the Patent granted to THOMAS HANCOCK, of Goswell Mews, Goswell Road, in the County of Middlesex, Waterproof Cloth Manufacturer, for Improvements in the Preparation or Manufacture of Caoutchouc in combination with other Substances, which Preparation or Manufacture is suitable for rendering Leather Cloth and other Fabrics Waterproof, and to various other purposes for which Caoutchouc is employed.

Sealed November 21, 1843. Enrolled May 21, 1844

FOR OBTAINING FORMS BY MOULDS AND VULCANISING. Specification of Patent granted to THOMAS HANCOCK, Stoke Newington, Middlesex, Esquire, for Improvements in the Manufacturing and Treating of Articles made of caoutchouc, either alone or in Combination with other Substances, and in the Means used or employed in their Manufacture.

Sealed March 18, 1846. Enrolled September 18, 1846

CONVERTING APPLICATIONS. Specification of the Patent granted to WILLIAM BROCKEDON, of Devonshire Street, Queen Square, Gentleman, and THOMAS HANCOCK, of Stoke Newington, Gentleman for Improvement in the Manufacture of Articles where India Rubber or Gutta Percha is used.

Sealed November 19, 1846. Enrolled May 19, 1847

PRINTING. Specification of a Patent granted to THOMAS HANCOCK, Stoke Newington, Middlesex, Esquire, for Improvements in Fabrics elasticated by Gutta Percha or any of the Varieties of Caoutchouc.

Dated November 2, 1847. Enrolled May 2, 1848

VULCANISED SOLUTIONS. Specifications of a Patent granted to THOMAS HANCOCK and REUBEN PHILLIPS for improvements in the Heating* or Manufacture of Gutta Percha or any of the Varieties of Caoutchouc.

Dated December 30, 1847. Enrolled June 30, 1848

* 'Heating' is a printer's error. In Thomas' copy of his private publication it has been crossed through and replaced with 'treating'.

APPENDIX II

Mechanical applications of vulcanized India rubber as described by Thomas Hancock

The list of applications below appears in both the 1857 and 1920 versions of Thomas' *Narrative* but is reproduced here to show the vast range of products which he conceived could be made of vulcanized rubber less than 15 years after the commercialization of the vulcanization process. The categories A, D, E, F and C have long ago vanished from any reference books and could well be specific to Thomas and his companies (J. L. Hancock and/or Charles Macintosh & Co.). Quality **A** has a density equivalent to 1.01 or 0.96 g/cm^3 depending on which of the quoted figures below one takes as a basis for the metric conversion. This indicates a gum rubber vulcanizate—that is, little has been added other than the chemicals required to effect a cure (in 1850s this would be only sulphur and possibly a metal oxide such as lead or zinc). **D** and **E** both contain inorganic materials as well as the curing chemicals which both reduce the elasticity and cheapen the product. As long as these are not added in a quantity which lowers the performance below that required this is perfectly acceptable and it can actually improve the service life beyond that of the gum vulcanizate—for instance it can impart abrasion resistance. Although carbon black is one of the most popular additives used today to improve the properties of rubber vulcanizates, it was not produced commercially for compounding with rubber until 1864 and so any earlier use would only have been of, say, soot to give a black colour to vulcanizates containing popular fillers such as talc or clays.

Mechanical Applications of Vulcanized India Rubber

NOTES.—**A** Quality is the most elastic. It weighs about 60 lbs. per cubic foot, or 1–29th of a lb. per cubic inch.

D Quality weighs about 82 lbs. per cubic foot, or 1/21th of a lb. per cubic inch.

E Quality more elastic than D—weighs about 92 lbs. per cubic foot, or 1/19th of a lb. per cubic inch.

F, C—Fibrous compound, used for flange washers, valves, and pump buckets. Weight 1/25th of a lb. per cubic inch.

THE following list describes some of the applications of india-rubber. Many of these articles are formed of pure vulcanized rubber, and others prepared with various pigments according to the required colour, quality, or intended application of the article, each modification of quality being distinguished by a letter, thus enabling the consumer to select either pure vulcanized rubber or any of the stated compounds.

BUFFER AND BEARING SPRINGS. (Fuller and D'Bergue's Patent.) A patent application of the vulcanized india-rubber to the purposes of draw springs, buffers, and bearing springs of railway carriages; more efficient, durable, and economical than any modification of steel springs for such purposes.

CYLINDERS. These are made of any dimensions of bore and thickness; they are supplied for the formation of buffers, washers, and springs, where enormous compression is used, and to relieve the concussion of steam hammers, fulling mills, &c. &c.

FOOT AND PUMP VALVES FOR OCEAN STEAMERS. These valves are rapidly superseding the metallic valves; concussion is avoided; a perfect joint formed under the most rapid motion of a steam engine; they are extensively used in steam ships constructed with paddle wheels; for screw steamers they are quite indispensable.

VALVE CANVASS. This is a modification of the above; it is prepared for various mechanical purposes, and is much less elastic than the vulcanized india-rubber, in which canvass or other fibrous materials are not incorporated.

ENGINE PACKING. This article, being a compound of india-rubber and fibrous materials, is eminently adapted for packing pistons, stuffing boxes, and the various parts of steam engines that require packing; it is supplied in sheets, slips or rings, and the fibres are so arranged in the compound as to give the greatest possible amount of durability and relief from friction.

WASHERS FOR FLANGE AND SOCKET JOINTS. By means of these washers an instantaneous joint may be made for every purpose, and in every conceivable situation in manufacturing establishments they are made to any figure or size, from that of the smallest pipe to the largest chemical cistern.

WHEEL TIRES. For this purpose the vulcanized india-rubber is firmly attached to a metal hoop or tire, of the usual width, or an endless band of rubber is sprung on to an ordinary wheel tire, and kept in position by a flange on either side; the rubber, projecting from the flanges, rests on the ground, and this prevents the concussion to which the wheels are ordinarily subjected, and altogether increases the durability of the carriage. Carriages having these tires roll along without the slightest noise, and in an extraordinarily soft and easy manner. They are particularly valuable for trucking goods in warehouses, railway stations, and for bath and invalid chairs.

ROLLING PISTON FOR LIFTING AND FORCING PUMPS. (Woodcock's Patent.) A patent application of the vulcanized india-rubber. The rolling pistons prevent all friction in pump barrels and water meters, and they are economical, enduring, and cannot "choke," whatever be the fluid they are employed to pump.

RINGS, STRIPS, AND CORDS FOR ELASTIC PURPOSES. These may be applied to the most delicate or most powerful mechanical purposes.

FLEXIBLE PUMPS FOR FORCING OR EXHAUSTING AIR, GASES, &c. &c. PUMP BUCKETS. These are made of various forms, and their advantages are, durability, and the facility with which water, liquid manures, and chemical liquids of any temperature may be pumped without injury to the vulcanized india-rubber.

PLUG VALVES. These are conical valves for ships, chemical and water cisterns, and as plugs for hot or cold water baths; they never corrode, or fail in forming a perfect joint.

COACH LOOPS OR ROUND ROBBINS. These are used by coach makers instead of those of combined iron and leather, as being more safe, quiet, and durable for the support of carriage bodies.

TEAGLE OR HOIST STRAPS. These are more enduring, safe, and economical than the ordinary hempen rope.

GAS BAGS. A useful and certain apparatus for the repair or alteration of gas mains; by its aid a town need no longer be placed in darkness during the laying of pipes.

DIAPHRAMS [sic]. These are made of vulcanized india-rubber, for dry gas meters, and for measuring the supply of water to towns, manufactories, or private dwellings.

DOOR SPRINGS. These are used in a variety of ways: one of its simplest applications is as a strong loop slipped over two hooks, one in the door and one in the jamb or door frame.

CORRUGATED RUBBER FELT. This article is extensively used for manufacturing and rail way purposes; in the latter, its peculiar properties are exhibited when placed under the chairs or flat rails; concussion is prevented, and wear and tear of the rails and carriages much diminished; it is also used extensively for the bottoms of vulcanized indiarubber over-shoes.

HOSE PIPES AND TUBING. These are made of any bore and length, suited to the delivery hose and suction pipes of fire engines, and ' for the conveyance of gas, steam, acids, alkalies [sic], and other fluids; they are also soft and pliable, and not injuriously affected by heat or cold.

LOCOMOTIVE PIPING. This is a modification of the above class of articles; for this purpose the ordinary strength is increased.

ROLLERS FOR LETTER-PRESS PRINTING. A substitute for the ordinary glue and treacle rollers. The chief recommendations are their permanent elasticity and durability.

BLANKETS FOR CALICO PRINTING. These are formed of alternate layers of india-rubber and cloth; by their use, a printer can produce fine and delicate patterns, that could not result from the ordinary woolen [sic] blanket. These patent blankets are capable of printing 25,000 pieces of cloth, and the power required is much diminished, in consequence of their peculiar elasticity requiring less pressure to produce the pattern.

FURNISHERS FOR CALICO PRINTING. These are formed of vulcanized india-rubber, with a roughened surface which takes up the colour, and applies it to the engraved roller. The same furnisher can readily be applied to any colour without waste; the composition of printing colours has no injurious effect on these furnishers, which perform their work for a series of years.

SIEVES OR FURNISHERS FOR SURFACE PRINTING. These are found to be a great improvement on the ordinary woollen sieve, especially in an economical point of view, as they prevent the absorption of the colour.

ARTIFICIAL LEATHER FOR CARD BACKS. This article is too well known to cotton and woolen manufacturers to need comment or explanation; suffice it to say, that its chief advantages are cheapness, evenness, large size of sheets, and elasticity.

MOULDED ARTICLES—ELASTIC. Under this head is comprised the endless variety of forms in which the vulcanized india-rubber is produced for manufacturing, surgical, domestic, and fancy purposes, which may require moulds for their production. It is the peculiar property of vulcanized india-rubber to maintain permanently the form in which it is vulcanized.

MOULDED ARTICLES-HARD VULCANIZED. Under this head is included all the before-mentioned; the degree of hardness is adapted to the purposes to which the substance is required to be applied. Hard vulcanized rubber is supplied in the form of sheets, slabs, bars, tubes, and can be moulded to any desired figure. It is a substitute for bone, ivory, whalebone, hardwoods, &c., and is capable of being worked by the ordinary tools used for those substances. It can be worked in the lathe, sawn, planed, drilled, screwed, or engraved. The properties of this material are: resistance to the action of hot and cold acids, alkalies or chemically impregnated solutions, the sulphur in gas; and it is suited to any uses in which metals are objectionable.

CUSHIONS FOR BILLIARD TABLES. These are an improvement upon the original india-rubber cushions, which in cold weather were hard, and in that state useless. The patent cushions never alter in their elasticity, and consequently are to be depended upon; they are highly approved of by scientific players.

SEWER AND SINK VALVES. (Dr. C. Bell's Patent.) These effectually prevent the escape of effluvium in all situations; they are simple in application, and the vulcanized india-rubber is not injured by contact with any fluids that may pass through the valve into the sewer.

CUMULATORS, OR ELASTIC POWER PURCHASES FOR PROJECTILE AND LIFTING APPARATUS, &c. (Hodge's Patent.) Here the vulcanized india-rubber is used in the form of tubes or cords. Simple or compound springs are brought to bear upon harpoons, arrows, balls, shot, and other missiles, and made to project them with immense velocity and precision. In case of lifting and suspending great weights, an assemblage of these springs are brought into use, by which a child may lift an enormous weight.

THREAD—VULCANIZED INDIA-RUBBER. This article, by reason of its great strength and permanent elasticity, has greatly extended the trade in elastic woven and knitted fabrics. It is prepared of several degrees of fineness, and supersedes the original native, or common, india-rubber thread.

VULCANIZED SHEET RUBBER. These sheets are supplied in grey or black of any thickness from the 70th of an inch upwards, and varying from 5 to 50 yards in length, by 50 inches and upwards in width; from these may be cut bandages, springs, strips for joints, and linings for chemical and other vessels, &c.

RUBBER AND CLOTH COMBINED IN SHEETS. This is commonly known as "insertion" rubber; where required, the cloth is made highly elastic, so as to stretch to the extent of the rubber; where this material is required not to stretch, a non-elastic cloth is used.

FINE CUT SHEET RUBBER. These sheets are supplied to any thickness, and are capable of being joined up for the manufacture of various articles.

FINE SHEET RUBBER. Made of any thickness from 36 inches wide and upwards, and from 5 to 50 yards long.

India-rubber Solution, or Varnish

Surgical Purposes

HYDROSTATIC BEDS. These beds afford great relief to the afflicted—facilitate their movement, and supply a soft support to every part of the person.

IRON BEDSTEADS WITH ELASTIC SACKING. These are adapted for hospitals and public institutions generally; they afford the greatest facility for attendance on the sick; are much cheaper than hydrostatic beds, and very simple in construction.

MATTRESSES, BEDS, AND PILLOWS. (With bellows for inflation.) These articles are prepared from air-proof materials, and when inflated assume their figure of mattress, bed, pillow, &c.; they are very easy, light, and portable, and made to any size and figure. Bellows are supplied for the inflation of the larger articles, and the operation is at once simple and easy.

ELASTIC WOVEN BANDAGES. Are formed as stockings, knee-caps, leggings, thigh pieces, anklets, armlets, wristlets, &c.; they are extensively used for the relief of glandular swellings, and varicose veins, abdominal belts, &c.; nothing can be more easy and efficacious in use, or more elegant in appearance, than these applications of the vulcanized india-rubber thread.

BED SHEETS. Applied as a cover, so that in cases of Hemorrhage, &c., a valuable bed is entirely protected from injury.

WATER PILLOWS AND BEDS. Are exceedingly elastic, either inflated for reclining on, or for the application of hot or cold water to any part of the body; to bed-ridden patients they are invaluable, as they entirely prevent the friction produced by ordinary cushions or pillows.

CHEMICAL APRONS, SLEEVES, AND GLOVES. Are used by surgeons in dissecting operations, arid prevent all risk from contact with poisonous fluids. They are also valuable for manufacturing chemists, dyers, and others, as they protect the person" and clothing from the action of caustic, alkalies, acids; and other dangerous liquids.

GAS VESSELS. Are made in the form of bags of any size or figure. They are used for the purposes of illumination, and for containing separate gases for chemical and experimental purposes; as, for instance, the oxy-hydrogen microscope, &c.

TUBING, BRAIDED OR PLAIN. For the conveyance of gas or other fluids, for moveable lights and general use in manufactories, chemical works, &c.

LIGATURES, &c. These are made in the forms of thread, cords, bandages, rings, &c., and are useful in cases" of dislocation, for the tourniquet and various other purposes in surgical operations.

INJECTION BOTTLES, BREAST BOTTLES, EHHAUSTING BELLS, ENEMAS, PESSARIES, URINALS, PIC NICS, EAR PADS, TRUSS PADS, CORN PROTECTORS, FINGER STALLS, &c. Appliances for medical and surgical purposes: their uses are indicated by their names, and are well adapted to their varied purposes.

Domestic Appliances

INFLATED CUSHIONS AND BEDS. Are made of any form and to any dimension, they are useful for chairs and sofas or for the purpose of traveling, particularly in second class carriages on the railway.

CHEST EXPANDERS. Afford agreeable and healthful exercise to children and persons engaged in sedentary employment, they strengthen the muscular power, expand the chest, and promote health. For schools and families they are particularly useful.

SPONGING BATHS. A portable and efficient accompaniment of the modern bed-chamber, no water is absorbed by the material, and the bath can be packed away immediately after use; hence its value to persons travelling.

JAR COVRRS AND CAPSULES. For pickles, preserves and anatomical specimens. They can be removed in a moment, and, soundly replaced, afford protection from the atmosphere, and are perfectly self-fastening

GUM RINGS, CORALS, NIPPLES, CRIB SHEETS, NURSING APRONS, ARM GUSSETS, SPONGE BAGS, STRAPS FOR BABY JUMPERS, DRESS DILATORS, BATHING CAPS, TOBACCO POUCHES, BOTTLING CORKS AND DUNGS, DECANTER STOPPERS, TABLE MATS, SEATS AND BACKS FOR CHAIRS AND STOOLS, PLAYING BALLS, FOOT BALLS, CRICKET GLOVES AND BAT COVERS, GLOVES OF ALL SIZES.

The names of the above articles indicate their use, and they are found safer and better adapted to their several uses than the articles they have superseded.

Wearing Apparel

PIECE GOODS, WATERPROOF FABRICS. CAMBRIC, SHEETING, LINEN, STUFF, ALPACA, SILK, WOOLLEN, &c. &c., DOUBLE AND SINGLE TEXTURE. CAPES. CAPES WITH SLEEVES OR LOOSE COATS. CHESTERFIELD WRAPPERS. COACHMEN'S COATS. LADIES' PALETOTS. BONNET HOODS WITH SHORT CAPE. OVERALLS. BRACES. VEST BACKS. GARTERS. GAITERS. TROUSER STRAPS. OVER SHOES. WEBBING FOR GUSSETS. WEBBING FOR BRACES. WEBBING FOR ELASTIC BOOTS. WEBBING FOR BANDAGES AND ROLLERS. SANDLING. APRON BANDS. WRISTLETS. LADIES' PAGES. ELASTIC BELTS. GLOVES.

Any explanation or statement of properties possessed by the above, is rendered almost superfluous by their general use by the public; but it may be observed that, of late years, the fashion has very much revived the use of waterproof clothing: the loose, open, easy character of the present dress renders a waterproof overcoat unobservable, made as they now are of materials such as are regularly worn in an un-proofed state, and consequently unobjectionable in appearance, whilst at the same time all the advantages of being kept perfectly dry in rain, is due to the india-rubber fabric alone.

Nautical and Agricultural Articles

SHIP SHEETS. Are used for the purpose of passing under a ship's bottom in case of leakage or accident at sea. They effectually stop the ingress of the water, and enable the ship to proceed on her voyage without repairing the leak; on arriving at a port, the damage can be effectually repaired without taking the ship into dock—advantages that have been appreciated by the Government and private traders.

SAFETY TUBES. (Holdsworth's Patent.) Are used for life boats, life buoys, watching buoys, &c., and are peculiarly adapted for giving buoyancy to boats of all descriptions. They can be placed fore and aft, secured by nettings to the raisings or rails fitted for the purpose, or be secured across the boats under the thwarts, as the judgment of the owner may direct, thus converting any boat into a life boat. As life buoys, they may be thrown to the assistance of persons falling overboard: they are so light as not to do injury to the person who may be struck with them, while their buoyancy is such, that 'they are capable of supporting three persons in the water until assistance arrives.

LIFE BELTS, OR LIFE PRESERVERS. Are made light, portable, and efficient for the protection of life and property in cases of wreck or other accidents at sea: many valuable lives have been saved by their use; and it is not too much to say, that in ordinary cases of shipwreck, the whole of the passengers and crew might be saved, if the properties of these valuable articles were more generally known.

BOATS INFLATED. Are intended for pleasure purposes on lakes, for fishing, and for exploring parties going abroad. They can be packed in small compass while travelling, and, when required, can be converted into a boat in five minutes.

SOU' WESTERS, DECK BOOTS, OVER SUITS, &c. Are essential to officers, and sea-faring persons generally, as they preserve the usual clothing perfectly dry in the roughest weather.

DIVING DRESSES. These entirely envelope the person, and enable divers and others engaged in submarine operations to perform their work in security.

CART WAGGON, AND RICK COVERS. Are thoroughly waterproof, very light, portable, and do not crack in use.

MALTING SHOES. Are used by persons engaged in malting, and similar operations. They can be attached to common shoes and enable persons to walk upon the grain without crushing or otherwise injuring it.

HOSE AND SUCTION PIPES. For the conveyance of liquid manure, and the general purposes of land irrigation.

HORSE STOCKINGS, BRUSH AND RING BOOTS, KNEE CAPS, AND SHOE PADS. These are found to answer the purpose much better than the ordinary leather ones, to which they are now preferred.

Travelling and Sporting Articles

TRAVELLING CUSHIONS, BEDS, RUGS, BAGS, &c., GIG APRONS, DRIVING GLOVES, CANTEENS AND BOTTLES, DRINKING CUPS, HORSE CLOTHS, MUD BOOTS, HANDLES FOR STICKS, UMBRELLAS, &c. SHOOTING BOOTS, GUN COVERS, GAME BAGS, SHOOTING HATS, RIDING BELTS, FISHING BOOTS, FISHING STOCKINGS, FISHING TROUSERS, FISHING COATS, YACHTING TROUSERS, TENTS.

Stationery Purposes

ELASTIC BANDS FOR PAPERS, LETTERS, &c. (Perry and Daft's Patent.),

INKSTANDS WITH ELASTIC BOTTLE, WRITING TABLETS, PARCEL BANDS, BOOK COVERS, ERASING RUBBER IN SQUARES AND BOTTLES.

Ornamental

ENAMELLED SHEETS, EMBOSSED SHEETS. MARBLED SHEETS, PRINTS FROM ENGRAVED PLATES, MAPS. BAS RELIEFS, MEDALLIONS, EMBOSSINGS, FLOWERS, FIGURES, ANIMALS, COLOURED THREAD, PLAIN AND BRAIDED.

Some of these articles exhibit the applicability of the vulcanized india-rubber to the Arts in cases where durable embossing of any degree of hardness, of fineness, of execution, susceptibility of colouring or elasticity are required. Among the applications of embossing are the production of durable books for the BLIND, and both hard and elastic type for printing.

APPENDIX III

The evolution of the rubber industry of today

Equipment and Machinery

Britain was lucky in that the industrial revolution had begun some years before Hancock started his experiments but, in terms of the size and mass of the equipment required to process rubber, it was still in its infancy. In part this was due to a lack of adequate power. In 1853 Goodyear wrote in his book *Gum Elastic*:

> It is for want of adequate power and corresponding machinery for this purpose, and of that only, that the inventor is dissatisfied with the present state of manufacture.

In the late eighteenth century the discovery that rubber could be dissolved in a suitable solvent to give a viscous syrup-like solution which could be painted on fabrics or used to dip formers in to make a particular product. This opened the door to a new industry, particularly since the equipment required for this was neither heavy nor had excessive power requirements. We have seen that although Hancock's 'pickle' (see page 22) was the first important piece of rubber manufacturing equipment, it was man-powered and produced but a few ounces of rubber per day. It could hardly be classed as part of the industrial revolution! The spreading machine (see page 54) which Thomas had built, probably from a design of Walter's, was little more than a large table with a trough to hold the rubber solution followed by an open drying oven.

Thomas Hancock was the first to introduce a power-driven machine into the rubber industry and this consisted of a single roller with a fluted surface rotating in a concentric case. As it rotated, the rubber was torn or masticated as in his 'pickling' machine and this could claim to be the forerunner of today's internal mixer. By the middle of the century ever-larger machines of the same basic design could handle up to 1800 pounds (900 kg) of rubber in one charge.

In America, in the 1830s, Edwin Chaffee concluded that the use of a solvent to prepare cast thin films of rubber was the main cause of the latter's rapid degradation so he developed a machine which he called a 'calender'. This had three or more rollers mounted vertically with adjustable 'nips' or gaps between them through which the rubber could be passed to produce ever thinner sheets. A sheet of cloth could be passed through the final nip with the rubber, enabling it to be rubber-impregnated without recourse to any solvent. A crucial point to notice is that the calender squeezes the rubber like an old-fashioned mangle; normally there is no shearing or mixing effect on the rubber between the rollers since they rotate at the same speed.

Chaffee also produced a two-roll, two-speed mill, which rapidly replaced Hancock's machines and which was specifically designed to masticate the rubber. When the pairs of rollers, mounted horizontally, rotated in opposite directions at slightly different speeds the rubber 'banded' on one roller while, because one roller was rotating more slowly than the other, the rubber bulked up at the nip and was masticated by the shearing action. Any additives that were required could be poured in between the rollers at the nip so that they became finely distributed throughout the mix. For that reason these mills were sometimes called mixing mills.

The calender and two-roll mill remain the basic equipment of virtually every rubber factory today and, although they may have changed with refinements such as heating or cooling the rollers and the fitting of safety guards, the fundamentals remain the same now as 150 years ago. The one major advance has been in the development of electrical power to drive them. The mills are manufactured in a vast range of roller lengths and diameters; a laboratory mixing mill might have rollers some 4 inches (10 cm) long and 1 inch (2.5 cm) diameter (see Fig. 12.9 Thomas Hancock's laboratory mill) while a large industrial mill might be 130 inches (2.5 m × 75 cm), able to take up to 4 hundredweight (200 kg) of final rubber compound and require a motor capable of delivering over 300 horsepower.

These mills had one particular disadvantage which was immediately evident when carbon black came to be added to rubber mixes in significant amounts. They were open to the atmosphere and the black got everywhere. The search for new methods of mixing which offered a cleaner working environment than the 'open' mill, and which was quicker and less power consuming, led to various 'internal mixers' being developed. Werner Pfleiderer, a company that had begun some 30 years earlier manufacturing machinery for mixing and kneading dough, made the earliest in 1913, while in 1916 Fernley Banbury launched the 'Banbury' mixer, which is probably the most popular mixer for rubber compounding around the world today. It uses a pair of counter-rotating winged tangential rotors in a machine which is fed from the top, and the final mixed material drops through doors at the bottom. Francis Shaw Ltd adopted the same principle in the 1930s with the 'Shaw Intermix', which still used rotors but these were designed in such a way that the materials being mixed were continually swept from side to side within the chamber to give consistently uniform mixed compound. Nevertheless, in spite of all the advances in mixing technology, it remains to this day a batch process, and no one has yet succeeded in devising a fully continuous mixing process.

The next stage in the manufacture of a vulcanized rubber product is to shape it and then apply heat to effect vulcanization. In 1855 Johnson patented the idea of using a press with shaped platens to form the product, and in 1860 Pitman introduced steam-heated platens. Some of these are still in use today, although electrically heated presses have generally replaced them throughout the last third of the twentieth century. While the final shape of the product obviously depends on the quality of the mould, in practice, the quality of the finished product can be crucially dependent on subtle details of the mould and the way in which the rubber compound flows to fill it. Mould design and manufacture is an extremely skilled job.

There are two basic types of moulding processes: compression moulding and transfer moulding. In the former a slab of compounded rubber is placed in the bottom portion of the mould and the top half is lowered on to it—thus compressing the slab to fill the mould—while in the latter, the compounded rubber is placed in a separate part of the mould (the transfer pot) which is connected to the mould proper by a series of small channels. A plunger forces the compound through the channels so that it transfers from the pot into the mould.

As with masticating and mixing the rubber compound, moulding with platens is a 'batch' process and this is obviously of little use in the manufacture of articles such as hoses or automotive rubber seals and profiles. In this area the rubber manufacturers learned from the plastics industry, specifically from Bewley, who had invented the plastics extruder in 1847 to extrude gutta percha as an insulating and protective coating for the first submarine telegraph cables (see page 103).

In 1881 Francis Shaw Ltd developed a screw extruder in which the compounded rubber was forced through an appropriately shaped die from a large cylindrical reservoir which was compressed by a screw piston. The product which was squeezed from the die (the extrudate) was then coiled on flat pans in spirals and cured in air ovens. More modern developments retain the extruder but use a variety of continuous curing procedures which, themselves, can operate a continuous cure.

Surprisingly it took over half a century for the extruder and shaped platens to be brought together in one machine. This process known is known as 'injection moulding'. Although it had been used by the plastics industry since the 1870s the process was not applied commercially to rubber until the 1940s, and it was only in 1957 that Arburg commenced series production of this type of machine. Early versions used either a simple ram or a screw ram to both inject the compound into the curing chamber and maintain pressure while the curing took place but in the 1960s screw machines with separate rams were introduced by Rep. These had a 'V' head design with one arm of the 'V' being the screw extruder which forced the compound through a non-return valve into a small chamber from which it was injected into the mould by a ram which constituted the other arm of the 'V'. This allowed a high degree of control over batch volumes and ram pressures, so minimizing wastage and improving product reliability. This type of machine is today fully automated and the industry standard.

This appendix has so far been concerned almost entirely with the heavy equipment needed to process raw rubber—natural or synthetic—but in 1920 Philip Schidrowitz discovered how to prevulcanize latex. His product looked and behaved exactly like virgin latex but when it was dried or coagulated and heated at a relatively low temperature, around 130°C, it gave a vulcanized product which had all the properties of conventionally vulcanized rubber. This led to the birth of a new industry and the manufacture of a vast range of dipped rubber goods, probably the three most important being surgical/domestic rubber gloves, condoms and balloons. Manufacture today takes place on dipping lines which run in a loop with tanks for dipping and then washing followed by a curing oven. The lines run continuously and are fully automated. Formers are made of wood, plastic, glass, or porcelain and these can be interchanged to meet demand, although it is usual for one line in a factory to keep to one product.

Prevulcanized latex can be sprayed on to fabric or expanded in moulds by means of a blowing agent so that vulcanized latex foam products, such as soft rubber balls or toys, can be made. It is also possible to whip latex into a foam and vulcanize to give, for instance, latex mattresses. This process was first developed by Dunlop in 1929, and in 1960 the Talalay process (named after its inventor, Leon Talalay) was patented in the UK.

Vulcanizing Chemicals

There can be no doubt that the discovery of the vulcanization process changed the world as much as, if not more than, any other discovery over the last few hundred years. In non-chemical terms it can be described as the process by which rubber and sulphur combine to give a product with greatly increased elastic properties and the maintenance of these properties over a comparatively wide temperature range.

Perhaps the most important—and surprising—thing to realize in the light of what has been written about vulcanization in previous chapters is that heating pure rubber and sulphur will not give a vulcanized product. Pure rubber in this context means just the polymeric material and this is not normally obtained by normal cleaning and 'purifying' in the factory. This material is only about 95 per cent 'pure polymer'. The reason natural rubber can be vulcanized with nothing but sulphur lies in the presence of natural proteins in the rubber which act as 'activators' to start the vulcanization process. Hancock's work showed that this was all that was needed, but Goodyear found that lead oxide also assisted the process. Unknown at the time, other natural materials present in the rubber, such as stearic acid, were also involved in the complex chemistry.

In 1881 Rowley patented ammonia as a promoter or activator for the vulcanization reaction but for obvious reasons this did not prove popular, and it was a further 20 years before anyone suggested using the less volatile organic base, aniline, as its replacement. Unfortunately its toxicity ruled it out very quickly and in 1911 Bayer suggested another base, piperidine, which was rejected for the same reason. In the early 1920s there were great advances with the discovery that two chemicals, diphenylguadine (DPG) and mercaptobenzthiazole (MBT) greatly assisted vulcanization. These types of chemical became known as accelerators, although in one sense the word is inappropriate because, although it did make the reaction proceed more rapidly, it also enabled the operator to have much more control of the vulcanization chemistry. Perhaps the origin of the word accelerator can be laid at the door of Peachy, who patented yet another amine in 1914 under the name 'accelerene'. Other accelerators were soon introduced, and in 1923 Bruni and Hopkinson, working in Italy, introduced a group of chemicals known as the dithiocarbamates. These accelerators, together with DPG and MBT, all of which were developed in the first half of the 1920s, remain the accelerators of choice for many applications today.

While Goodyear and Hancock had shown that vulcanization worked in a practical sense, there was a period of close to a century after their discoveries when the argument raged about what the interaction between rubber and sulphur actually was. In 1898 Ostromislenski proposed a combined chemical/physical theory of vulcanization, in 1902 Weber proposed a purely chemical one, and in 1910 Ostwald opted for some sort of physical mixture or 'alloy' formation. Although the evidence for a chemical inter-reaction became overwhelming, for most scientists there were still arguments between the various camps at an international conference on vulcanization as late as 1939 and it was only in the 1970s that research finally showed in detail how sulphur and rubber chemically reacted to give the material we call a rubber vulcanizate.

Chemicals to Reduce Degradation of Rubber Products

Now that vulcanized rubber products could be manufactured it was soon obvious that there were forces at work which degraded them after a relatively short period of time. Hancock, in his *Narrative*, repeated the observation which he had first made about 1826:

> The injurious effects of the sun's rays upon thin films of rubber we discovered and provided against before much damage accrued.

He had obviously realized at a very early stage that sunlight induced degradation of rubber, but in spite of his assertion above it was certainly not 'provided against'! What Hancock did not realize was that it was not the light which degraded the rubber but light-catalysed oxidative degradation, something shown by Spiller in 1865.

The story of the understanding of rubber degradation and the ways of limiting it, albeit only to a certain extent, is very much one of the twentieth century. In 1895 Henriques discovered that if a piece of natural rubber was extracted with acetone and then vulcanized it showed a much greater tendency to oxidize than did the rubber if it had not been extracted, but it was only in 1931 that Dufraisse and Drisch showed the converse—that if the extract from a sample of natural rubber was recombined with a previously extracted rubber, much of the protection returned. Their conclusion was that there must be some naturally occurring chemicals present in the 'raw' rubber which acted as antioxidants.

At the turn of the nineteenth–twentieth centuries it was discovered that amines and amine-based materials offered considerable protection against oxidative degradation, but it took until about 1930 for purpose-designed chemicals intended to reduce the damage brought about by oxygen (and ozone) to become commercially available. These were known collectively as amine-based antioxidants and antiozonants. However, they had one considerable handicap: although they were pale straw in colour when first synthesized, they themselves soon oxidized to various shades of blue, purple, and black. While this might not be too important in a black vulcanizate such as a car tyre or an industrial bearing, it was certainly of little use in the many light-coloured products such as rubber thread or mackintoshes.

Today the same group of chemicals, called para-phenylenediamines (PPDs), are regularly used and function as both antiozonants and antioxidants. As antiozonants they operate by reacting more readily than the rubber with any ozone present at the surface of the article. In so doing they build up a protective film which, as it thickens by migration and further reaction of the migrated antiozonant, eventually provides an impermeable barrier to the gas. Since part of the mechanism by which these antidegradants operate is by migrating to the product surface, they could then further migrate into any material in contact with the rubber and oxidize in their new environment, so producing a dark stain. There are many different PPDs and one of the main reasons for a manufacturer to select a particular one for a specific product is its solubility and rate of migration in the polymer system being protected. Since this is so obviously a crucial factor in the long-term protection of a product, it is not unusual for several such materials to be added and so give a broad spectrum or extended lifetime of protection. However, because they so readily stain, their use must obviously be selective.

It took a further generation (the late 1940s/early 1950s) for 'non-staining' antioxidants based on phenols to be developed and brought to the market, although Murphy had patented the use of phenol and cresol as antioxidants as early as 1870. Everything is relative and it was soon realized that while these antioxidants were certainly 'non-staining' compared with many of the amine-based materials, it was not true to say that the rubber would remain colourless, or that it might not discolour materials in contact with it. Indeed, pronounced yellowing of fabrics containing rubber thread so protected was not uncommon.

Phenolic antidegradants are not considered antiozonants, only antioxidants. When oxygen attacks a rubber molecule there is the potential for several chemical reactions to occur and the phenols can only offer an alternate chemical pathway which disrupts part of the degradation process. It is therefore essential to realize that they do not stop the oxidation outright and their effect is, at best, to slow down the oxidative breakdown. They do not form a protective skin if oxidized on the rubber surface, indeed, because of their mode of operation they need to be intimately dispersed or dissolved in the rubber to function.

The yellow discolouration mentioned earlier of uncoloured vulcanizates may sometimes be related to poor quality or inappropriate grades of white filler or brightener (although this hardly ever results in the staining of contact fabrics). There are, however, certain cases where some yellowing has been traced to the phenolic antioxidants themselves. This was initially put down to impurities in the chemicals but it soon became apparent that it was oxidative degradation of the antioxidant that was producing a derivative which was coloured yellow. The yellowing effect was found to be stronger in urban or industrial environments than rural ones, suggesting that the yellowing could be related to the oxides of nitrogen, which are nowadays major atmospheric pollutants, produced as by-products in the high temperature combustion of fuels by motor vehicles, industrial boilers, etc. Research in the 1990s identified these yellow impurities and showed how the chemical structure of phenolic antioxidants could be selected or modified so that this chemistry did not occur and the yellowing could therefore be eliminated.

Inorganic Fillers

With the exception of dipped goods, it is usual for most commercial rubber products to contain appreciable levels of other materials. Inorganic powders have been used to dust rubber surfaces, and so reduce their stickiness, since the first Mesoamericans used rubber to fabricate bouncing balls. Initially they would just have used dried powdered earth, and little changed until the nineteenth century when Thomas Hancock began to study the effect of a range of inorganic chemicals on rubber surfaces. However, it was not until he invented his masticator and Edwin Chaffee developed his mixing mill that it became relatively easy to disperse these powders uniformly in a rubber matrix.

Hancock's patent for vulcanizing rubber has already been mentioned, and the point made that the title contains no specific reference to that process.

In fact the long patent, running to over three 3000 words, is at least half concerned with the incorporation of magnesium silicate (talc) and other inorganic fillers, as well as powdered asphalte, into rubber (see Appendix I). In Goodyear's book of 1855 he mentions using magnesia, lime, and white lead as well as various colorants such as chrome salts, while by the turn of the century one firm was advertising antimony salts, barytes, black pigment, cadmium yellow, white flake, French chalk, various clays, red lead, lime, litharge, lead oxide, and zinc oxide, a reasonably comprehensive list even by today's standards.

With the discovery of vulcanization and the realization that rubber products could actually have reasonable life expectancies, the demand for them began to grow and, with the raw material being in short supply and therefore expensive, anything which could be done to bulk them out with cheaper materials which did not detract too much from their expected properties was welcome. However, manufacturers were yet to appreciate that certain fillers could actually improve the properties of a vulcanizate. It was left to Heinzerling and Pahl, in 1891, to carry out a classic experiment in which they compared the properties of a range of vulcanizates which were identical in every way except that they contained different inorganic fillers. They were able to reach certain conclusions which had a degree of validity but, not knowing that a number of fillers actually had an effect on the vulcanization process, some accelerating it and others retarding it, the 'identical vulcanization conditions' did not bring about chemically identical vulcanizates and their conclusions were not always valid. However, they were certainly correct when they reported that zinc or magnesium oxide 'strengthened' the vulcanizate more than a filler such as silica. It also became obvious that the quality of the fillers was sometimes suspect, and that certain impurities such as copper, manganese, nickel, chromium, and cobalt salts could have a very deleterious effect on the longevity of a vulcanized product.

By the 1920s most of the fillers in use today—talc, chalk, clays, and barytes—had been evaluated and adopted by the industry, although one relatively modern material, titanium dioxide, only started to be commonly available in the early 1950s. This white powder had, and continues to have, two significant advantages over filler levels of zinc oxide, although not replacing it as a vulcanization activator. It has a much higher tinting strength, i.e. it is a 'brighter' white pigment and it lacks any toxic properties.

With the advent of the motor car came demands for a tyre which would last more than a few months, and this required a 'strengthening' or reinforcing of the rubber way beyond that which then could be obtained by the use of the various inorganic fillers. Carbon black was about to come into its own. This material had been known for centuries and was used by the Chinese and Egyptians to make inks and eye make-up. They burned resins, fats, and oils under inverted pottery cones and collected the soot which became deposited on the pottery surface. Today this soot is called 'lamp black'. Both Goodyear and Hancock had used this as a colorant and at filler levels without noticing its reinforcing or strengthening properties. A variation on the lamp black process, manufacturing what became known as 'channel black', was invented by McNutt in 1892, in the United States, and this also found its market in the ink and printing ink business as well as in the rubber industry.

In 1895 the first motor vehicle specifically designed to run on pneumatic tyres took part in the Paris to Bordeaux (and back) race. In spite of accumulating 22 punctures it finished a creditable ninth from a field of 42. The era of air-filled, or pneumatic, tyres had dawned, and by the end of that century most of the Western tyre manufacturers who are household names today were manufacturing pneumatic tyres: Dunlop (1889), Michelin (1895) Goodrich (1896), Goodyear (1898), Firestone (1900).

As we have seen (page 200), in 1904 S. C. Moke, working at the India Rubber, Gutta Percha & Telegraph Works at Silvertown, showed how carbon black could be used to give a remarkable increase in the mechanical strength of a rubber vulcanizate, but it was a number of years before this was adopted by the tyre industry in general, probably only when tread wear took over from tyre fabric wear as the major cause of short tyre lives.

In 1916 Brownlee and Uhlinger introduced a new process in which gas was combusted to carbon black by injecting it directly into very hot ovens, and they found that by changing conditions a wide range of particle sizes could be obtained. These were called 'thermal blacks', and this type included the largest of all the black particles. A third process was developed in 1922 by the Columbian Carbon Company that again burnt gas but this time in a furnace, giving rise to their designation as 'furnace blacks', while the combustion of acetylene was found to give yet another type, 'acetylene black', which found a use in the manufacture of electrically conductive rubber.

In 1923 Frank and Marckwald prepared vulcanizates which were identical in all respects except that one contained German lamp black and the other, American oil black. They found that their physical properties were significantly different, with the former giving a more elastic product and the latter a much tougher material. It was realized that carbon blacks produced by different routes and from different starting materials gave a range of products that differed not only in particle size but also in structure and purity. By varying the type of black used in an otherwise consistent mix it was possible to impart a wide range of different physical properties to a rubber product and today blacks are available in many grades, the correct choice of which is of crucial importance to the performance of the finished product.

Although it has been said that fillers were originally added to bulk out a formulation no mention has been made of what level constitutes a 'filler'. Small additions make no real contribution to quality or cost, although they may on occasion be used to fine-tune the properties of a vulcanizate so that it meets some specification. Because of the complexity of rubber mixes, levels of added ingredients are always quoted as the number of parts by weight per 100 parts of rubber (that is phr); inorganic fillers were, and still are, added at levels typically between 25 phr and 150 phr, although in certain applications, such as rubberized carpet backings where the elastic properties are utilized to a minimum, levels of up to 600 phr have been seen. Goodyear mixed magnesium oxide with rubber to such an extent that he was able to make buttons and knife handles from the stiffened product. Titanium dioxide has been mentioned as a filler but it is expensive, and when white goods are required it is commonly used at relatively low levels, perhaps 5–10 phr, as a 'brightener' in conjunction with another 'bulk' filler such as chalk or talc.

Carbon black can be used just as a colorant at levels of 1 phr or so, but for reinforcing purposes it normally falls in the same range as the inorganic fillers, 25–150 phr. As the black loading of the rubber increases, so does the stiffness of the final vulcanizate. To counteract this effect, oil can be added, thus reducing the rubber content even further. A 'play-off' of black and oil loadings can be used to manipulate the performance and durability of a particular product.

Given the propensity of rubber product manufacturers to add almost anything available to the basic rubber mix in the hope of producing some particular advantage, it is not surprising that as soon as synthetic elastomers became available they too were mixed with natural rubber to give a blend of elastomers, and today the elastomeric component of many products is such a blend. Moving from the blending of different elastomers to the blending of an elastomer with a non-elastic polymer (a plastic) seemed an obvious next step and this resulted in a new class of materials, thermoplastic elastomers. These perform as one would expect from a vulcanized elastomer at ambient or near-ambient temperatures, but, unlike a vulcanizate, they can be remoulded when heated above the melting temperature of the plastic. Surprisingly, such a material was first made by Thomas Hancock in 1848 when he mixed natural rubber with gutta percha, but it seemed to have no commercial use and the concept lay dormant for over 100 years until the advent of modern synthetic plastics.

The rebirth began in the 1960s in the plastics, rather than the rubber, industry when low levels rubbery materials were blended with the plastic polypropylene to overcome the low temperature brittleness of the latter. These were called 'impact modified plastics'. Soon the whole ratio range was investigated and at the other end of the scale a polypropylene-modified synthetic rubber was patented in the early 1970s. The use of natural rubber in similar blends also began in the 1970s and this type of thermoplastic elastomer is available commercially today. It is not surprising that these and similar materials are making considerable inroads into certain areas which were previously the prerogative of the vulcanized material as they offer considerable advantages in recycling and waste management—not least the absence of sulphur. The main limitation is their restricted operating temperature range which, for practical purposes, cannot be considered to be much above 'very hot ambient'.

Ebonite

In previous chapters mention has been made of ebonite—the product obtained by heating rubber with a relatively large amount of sulphur. Because of its unique importance it is appropriate to expand here on the history of the material.

The levels of sulphur used to vulcanize rubber rarely rise above 3 phr, but when it is added at considerably higher levels, typically 25–50 phr, a quite different material is produced and that is ebonite. This hard material is known by a number of names including vulcanite, preferred by collectors, and hard rubber, preferred in the United States.

Its origins are obscure. Various eighteenth-century scientists prepared hardened rubber, but it seems likely that these were mostly the heavily oxidized material. The first record of rubber possibly reacting with a sulphur-containing chemical was made by Roxburgh in 1801, who obtained a white inelastic material when he passed chlorine into a solution of rubber in carbon disulphide and then poured the solution into water, but there are no analytical data to show that this was ebonite. In 1831 Leuchs added sulphur to molten rubber and got a violent reaction which resulted in a coal-like hard mass, which might just have been ebonite but he did not pursue the chemistry further.

Nathaniel Hayward introduced Charles Goodyear to sulphur when he showed that dusting rubber with sulphur and exposing it to sunlight (the 'solarization' process) gave it a hard skin, and this could well have been an ebonite skin. In fact Ludersdorff had done exactly the same thing some years previously but not in Goodyear's presence! The first authentic samples of ebonite were prepared by Thomas Hancock some time prior to 1843, and the details are given in his 'vulcanization' patent. By 1846 he had obtained the first patent for its use, and that was for making moulds for the vulcanization of ordinary 'soft' rubber vulcanizates (see Appendix 1). In the United States Charles Goodyear and his brother, Nelson, were heavily into ebonite, although the first US patent was only granted to Nelson in 1851. In the same year he took out British patent 13542 for the manufacture of ebonite by mixing rubber, 50 phr sulphur, and 50 phr mineral filler and then heating for between two and six hours at temperatures around 270°F (132°C). In the patent he listed applications such as buttons, door knobs, and inkstands. Between 1851 and 1855 Charles Goodyear took out a further nineteen patents relating to ebonite, listing its application in the manufacture of a wide range of products. He described his range of ebonites:

> The hardest of these compounds resembles marble; that which is less hard, ivory and buck-horn; that is still softer, buffalo-horn and whalebone; while they possess, in general, more durable properties than any of the substances above named, except marble, and they are even more substantial than that, in some respects; because, in all degrees of hardness, they have a great degree of toughness or tenacity, and the property of retaining the shape into which they have been moulded and heated.

Hancock was less passionate about ebonite but still appreciated it and used it extensively:

> The hard vulcanised rubber has been applied to many useful purposes to which this patent has contributed. Combs, knife and other handles, ornamental panels for carriages and furniture, stop-cocks, tubing, pump-barrels, pistons and valves for use in chemical works, &c. &c.,—these are capable of being turned in the lathe, and to have screws cut on them in the same manner as is practised with wood, ivory or metal. I have also had some flutes made of it, the colour is a jet black, and it polishes like ebony; the notes or sounds are equal to the best flutes, whilst they are said to be produced with greater ease by the performer. I furnished the material to the flute maker without instruction, and he made it in his ordinary practice... We have supplied it by the ton for the use of comb-makers, who like it, not only because it makes a good saleable article, but because they can have it in large sheets of the thickness they require, and make much less waste than when using such small pieces as are produced in horn, tortoise-shell, &c. The turner, the engraver, the comb-maker, and most other artists and mechanics, have only to apply their ordinary means, tools, and skill as to wood, ivory, metal, and other substances. It is also a fair substitute for whalebone and walking-sticks, and also more delicate articles, as bracelets, gold and silver mountings, pens, and penholders, picture-frames; and one may go from these to the contrary extreme, and if it were economical, or in any way advantageous to do so, it would make good houses, ships, wagons, and carts, and almost everything where wood is now employed, which I only mention to show the universality of its application, and in general, by the ordinary means practised in the different departments of Art.

Because of its ease of fabrication and its usefulness in a pre-synthetic plastics age, demand was high. In the UK Charles Macintosh and Co. had already manufactured articles for the 1851 exhibition and continued to do so. The India Rubber, Gutta Percha and Telegraph Works moved into this field in 1860, and in 1861 the Scottish Vulcanite Company became the first British company to be formed solely for the purpose of manufacturing ebonite articles.

In the United States there was similar activity with several firms making ebonite products in the early 1850s. One of these was Poppenhusen and Koenig, which was formed in 1852 after Conrad Poppenhusen split with his long-term partner, Adolph Meyer, with whom he had been manufacturing combs and corset stays from a diminishing supply of whalebone. Soon after, Meyer returned to his native Hamburg to found Harburger-Gummi-Kamm-Co. in 1856, which manufactured ebonite combs.

Apart from domestic uses, ebonite was in great demand from industry as an inert and electrically insulating material. Ebonite pumps had been manufactured in the 1860s but by 1900 it was possible to construct a complete chemical manufacturing plant from ebonite. It was also used in telegraph and later radio equipment as well as the ubiquitous car battery case which survived until relatively recently.

In the United States there was similar activity with several firms starting to [illegible] in the early 1970s. One of these was Computervision and Boeing [illegible]

[illegible]

Bibliography

COURT CASE

Bewley v *Hancock*. (1855–1860)—transcription of the proceedings in Chancery.

JOURNALS

India Rubber Journal, first published in the UK in 1884. It later became the *Rubber Journal* and then the *European Rubber Journal.*

BOOKS

Alexander, J.H. (1967) *More than Notion*. Fauconberg Press, London.

Anon. (1942) *Thomas Hancock Nunn—The Life and Work of a Social Reformer.* Baines and Scarsbrook Ltd, London.

Bee, M. (1997) *Industrial Revolution and Social Reform in the Manchester Region*. Neil Richardson, Manchester.

Cantor, G. (1991) *Michael Faraday.* Macmillan, London.

de Chasseloup Laubat, F. (1942*) François Fresneau, Seigneur de la Gataudiére, Père de Caoutchouc.* Les Petits-fils de Plon et Nourrit, Paris.

Gibbs, R. (1885) *History of Aylesbury.* Bucks Advertiser and Aylesbury News Office, Aylesbury.

Goodyear, C. (1855*) Gum Elastic.* The India-Rubber Journal's Facsimile Edition. (1937). Maclaren & Sons Ltd: London.

Hancock, T. (1853) *Specifications of Fourteen Patents for the Treatment and Application of India Rubber.* George Barclay. London (for private circulation).

—— (1857) *Personal Narrative of the Origin & Progress of the Caoutchouc or India-rubber Manufacture in England.* Longman, Brown, Green, Longmans & Roberts, London.

Hancock, W. (1838) *Narrative of Twelve Years' Experiments (1824–1836) Demonstration of the Practicability and Advantage of Steam Carriages on Common Roads.* John Weale, High Holborn.

James, F. (1975) *Walter Hancock and his Common Road Steam Carriages.* Laurence Oxley, Alresford.

Loadman, J. (2005) *Tears of the Tree: the Story of Rubber—A Modern Marvel.* Oxford University Press, Oxford.

Love and Barton (1839) *Manchester As It Is.* Love and Barton, Manchester.

Macintosh, G. (1847) *Biographical Memoir of the late Charles Macintosh, F.R.S.* W.G Blackie, Glasgow (for private circulation).

Schidrowitz, P. and Dawson T.R. eds (1952) *History of the Rubber Industry.* W Heffer & Sons Ltd, Cambridge.

Spencer, George Moulton & Co. Ltd (1948) *1848—1948 A Hundred Years of Rubber Manufacture.* Privately published by George Spencer Moulton & Co., Bradford-on-Avon.

Pierce, B.K. (1866) *Trials of an Inventor: Life and Discoveries of Charles Goodyear.* Carlton & Porter, New York.

Woodruff, W. (1958) *The Rise of the British Rubber Industry During the Nineteenth Century.* Liverpool University Press, Liverpool.

Index